DARWIN

Charles Darwin was one of the most influential scholars of his time. His ideas continue to have a far-reaching impact and influence on modern thought in the arts and on society, as well as in science. With contributions from leading scholars and respected communicators, this collection of essays is written for the general reader but from the standpoint of the leading researcher. They explore how Darwin's work grew out of the ideas of his time, and how its influence spread to contemporary thinking about the limits of human evolution and the diversification of living species and their conservation. The book provides a full account of the legacy of Darwin in contemporary scholarship and thought.

Featuring contributions from Janet Browne, Jim Secord, Rebecca Stott, Paul Seabright, Steve Jones, Sean Carroll, Craig Moritz, Ana Carolina Carnaval and John Dupré, this book derives from a highly successful series of public lectures.

WILLIAM BROWN is the Master of Darwin College and Professor of Industrial Relations in the Economics Faculty at Cambridge University. His research has been concerned with collective bargaining, pay determination, incomes policy, payment systems, arbitration, minimum wages and the impact of legislative change. In 2002 he was awarded a CBE for services to employment relations.

ANDREW C. FABIAN is the Vice-Master of Darwin College and Royal Society Professor of Astronomy at the Institute of Astronomy in the University of Cambridge. His research interests centre on black holes and clusters of galaxies. He has organised several previous Darwin Lecture Series (*Origins* in 1986, *Evolution* in 1995 and *Conflict*, with Martin Jones, in 2005). He is a Fellow of the Royal Society and was awarded an OBE in 2006.

In the second term of every academic year since 1986, Darwin College, Cambridge has organised a series of eight public lectures. Each series has been built around a single theme, approached in a multidisciplinary way, with each lecture prepared for a general audience by a leading authority on his or her subject. Collections of these lectures are published by Cambridge University Press.

Subjects covered in the series include:

23 DARWIN
   eds. William Brown and Andrew Fabian
   pb 9780521131957

22 SERENDIPITY
   eds. Mark de Rond and Iain Morley
   pb 9780521181815

21 IDENTITY
   eds. Giselle Walker and Elizabeth Leedham-Green
   pb 9780521897266

20 SURVIVAL
   ed. Emily Shuckburgh
   pb 9780521718206

19 CONFLICT
   eds. Martin Jones and Andrew Fabian
   hb 9780521839600

18 EVIDENCE
   eds. Andrew Bell, John Swenson-Wright and Karin Tybjerg
   pb 9780521710190

17 DNA: CHANGING SCIENCE AND SOCIETY
   ed. Torsten Krude
   hb 9780521823784

16 POWER
   eds. Alan Blackwell and David Mackay
   hb 9780521823717

15 SPACE
   eds. Francois Penz, Gregory Radick and Robert Howell
   hb 9780521823760

14 TIME
   ed. Katinka Ridderbos
   hb 9780521782937

THE DARWIN COLLEGE LECTURES

DARWIN

Edited by *William Brown* and *Andrew C. Fabian*

CAMBRIDGE
UNIVERSITY PRESS

CAMBRIDGE UNIVERSITY PRESS
Cambridge, New York, Melbourne, Madrid, Cape Town, Singapore,
São Paulo, Delhi, Dubai, Tokyo, Mexico City

Cambridge University Press
The Edinburgh Building, Cambridge CB2 8RU, UK

Published in the United States of America by Cambridge University Press, New York

www.cambridge.org
Information on this title: www.cambridge.org/9780521131957

First published 2010

A catalogue record for this publication is available from the British Library

Library of Congress Cataloguing in Publication Data

Darwin / edited by William Brown and A. C. Fabian.
    p. cm. – (Darwin College lectures ; 23)
ISBN 978-0-521-13195-7 (pbk.)
1. Darwin, Charles, 1809–1882–Influence. 2. Evolution (Biology) I. Brown, William Arthur.
II. Fabian, A. C., 1948– III. Title. IV. Series.
  QH366.2.D34184 2010
  576.802092–dc22

                                                    2010010000

ISBN 978-0-521-13195-7 Paperback

Contents

Contents

Contributors

Janet Browne Department of the History of Science, Harvard University, Science Centre 371, Cambridge, Massachusetts, USA

Jim A. Secord Department of History and Philosophy of Science, University of Cambridge, Free School Lane, Cambridge, UK

Rebecca Stott Faculty of Arts and Humanities, University of East Anglia, Norwich, UK

Paul Seabright Toulouse School of Economics, Institut d'Economie Industrielle, Manufacture de Tabacs, 21 Allee de Brienne, Toulouse, France

Steve Jones Research Department of Genetics, Evolution and Environment, University College London, Gower Street, London, UK

Sean B. Carroll R M Bock Laboratories, University of Wisconsin-Madison, 1525 Linden Drive, Madison, Wisconsin, USA

Craig Moritz Museum of Vertebrate Zoology, University of California, Berkeley, Valley Life Science Building, #4151, Berkeley, California, USA

Ana Carolina Carnaval Committee on Evolutionary Biology, University of Chicago, Illinois, USA

John Dupré Professor of Philosophy of Science and Director, ESRC Centre for Genomics in Society (Egenis), Byrne House, St German's Road, University of Exeter, UK

Introduction

Two hundred years after his birth, Charles Darwin's reputation has never been higher. His influence upon both biological science and our broader understanding of humanity continues to be profound. In this book, leading authorities in different fields are brought together to evaluate this influence, how it has developed and how our understanding has moved further. Their contributions originated in the 24th annual series of public lectures organised by Darwin College, Cambridge, a series which has always called on leading scholars, across the range of disciplines, with outstanding reputations as popularisers. It is, accordingly, the purpose of this book to provide a broad overview of current thinking on Darwin's influence, and one which will be widely accessible.

We start with Darwin's impact upon his own world, and how this was to echo through the wider realms of arts, literature and the social sciences, each responding to the reflected insights from different facets of his work. The story then moves to biological science itself, now able to deploy analytical tools unimaginable 200 years ago. What can we say about human evolution? What does modern science say about the evolutionary process and the implications of environmental degradation? And where does subsequent science put Darwin's theory? Is it as robust as when it was conceived, or has it been augmented by complementary or competing theories of origins of species?

Few thinkers of the modern age can compete with the magisterial popular image of Charles Darwin. Janet Browne starts the book, exploring the making of this image in an essay on biography. She shows how successive generations have, according to their changing concerns, reinterpreted Darwin as safe and domesticated, or as a tormented iconoclast, or as an archetypal biologist. And, most recently, we have come to see him as a supreme networker, absorbing and combining the emerging ideas of his age through tireless use of the penny post.

The potency of Darwin's incessant letter writing is taken up by James Secord, who shows how he used it to focus a variety of widely discussed, if inchoate, theories of evolution. By mobilising a great array of widely collected facts, Darwin did not so much produce a new theory as legitimise both an emerging theory and, at least as important, a way of doing science. Thus focused, the debate he provoked was to have a catalytic effect on different local debates among scientists right around the world.

Nor was the influence confined to science. The notion that living things were not created, once and for all time, in their present form, but that they have changed and continue to change form in response to environmental and other pressures, was deeply subversive. It was to have a profound and recurring impact on humanity's view of itself as well as of the natural world. Rebecca Stott explores how this has influenced literary fiction, an art form of which Darwin was particularly fond. Optimistic views of mankind as the pinnacle of creation could hardly survive the horrors of the twentieth century. By drawing our attention to the elegiac quality of much of Darwin's prose, she emphasises the tangle of delight and terror with which he himself viewed his conception of life's central process, where death is the inextricable companion to natural beauty.

Contrary to popular opinion, cooperation was as important as competition for Darwin. How this has shaped our understanding of human society is explored by Paul Seabright. Drawing on studies of primates and primitive human communities, he shows how collaboration is itself central to the competitive process. Humans developed as they did because there were substantial survival benefits to be gained from complex cooperative behaviour. The development of institutions and of cultures was accompanied by a steady decline in the incidence of deliberate, violent death. The consequent importance of sexual selection for humans makes it particularly notable that it is so difficult to find significant differences in overall cognitive abilities between men and women. The implication is that, as our species' distinctively complex cooperative behaviour evolved, the genders were facing equally demanding cognitive challenges.

This brings us to the question of whether we, as humans, are continuing to evolve. Steve Jones tackles this question from his background in genetics. For Darwin, the central characteristic of living things is that they make copies of themselves. But they do so with small errors, and their continuation through natural selection feeds on the improbability that small, inheritable variations will fit changing environmental circumstances. Civilisation tends to stifle this process. People increasingly die old, not young, having passed on their DNA. There is less variance in family size, in personal survival and in reproductive success. The variation associated with inbreeding has become less likely because we have changed from being an often endangered species to one that has become, relative to other species of comparable body size, enormously abundant, and also more mobile. If humans are going to get better, it will have to be by using their brains. So far as natural selection alone is concerned, barring disasters, we are probably as good as we are going to get.

Part of the challenge facing Darwin in getting his ideas accepted was that people have, inevitably, a deeply parochial awareness of the natural world. Evolutionary processes that involve geological time and molecular distances are beyond our comprehension. The notion of a constant transformation of species across time and space challenges our daily experience. However, modern molecular biology has provided the means to explore these elusive dimensions. As a result, evolutionary research has moved from the collecting and analysing of species of Darwin's day to collecting and analysing their DNA sequences.

The opportunity to look far back in time is offered by the curious fact that the redundant genetic codes of past forms are to be found within species' living DNA sequences. Sean Carroll discusses how these biological fossils permit the exploration of the evolutionary history of species. They show how species have adapted to changing environments. A constant 'arms race' of evolution changes both the hunters and the hunted. The frequent repetition of evolution shows how similar selective conditions favour similar genetic variations in different species, at different times, in different parts of the world. Modern species are not 'better' than their ancestors; they are just different.

The importance of spatial separation for evolutionary processes was evident to both Darwin and his great contemporary, Alfred Russel

Wallace, who were famously inspired by the distinctive fauna of islands. Biogeography has subsequently become a central part of their legacy. Craig Moritz and Ana Carolina Carnaval discuss how the mapping of biodiversity has developed. However, their main concern is with the processes whereby biodiversity is first generated, and then preserved. Newly evolved species are associated with complex topographies, steep environmental gradients, novel environments and recent climatic stability. But the dominant factor influencing the preservation of species appears to be the stability of overall climate conditions. As the planet faces hastening climate change, biodiversity prediction is central to effective conservation policy.

Our discussion of Darwin's heritage ends by questioning the dominance of the theory based on the combination of Darwin's natural selection and Mendel's genetics. John Dupré argues that science has taken us much further. For the most numerous of organisms, microbes, lateral gene transfer provides no lineal descendants. The tree of life merges as well as branches. And for more complex bodies, symbiosis has taken forms in which distinct organisms have bonded in cooperation to create new life forms. Evolution has often been by merger. The definition of an organism is itself an issue. Might it not be more useful to envisage the typical organism as a collection of cells of different kinds, organised cooperatively to maintain its structure and reproduce similar structures? And if that is so, might it not be appropriate to focus not on a single genome, but on the life cycle of the self-copying multigenomic whole? In seeing evolution as a mosaic of more or less related processes, we do not detract from Darwin's achievement, but rather acknowledge the fertility of the intellectual revolution associated with his work.

There is no doubt that Charles Darwin was a superb scientist. He spent so much of his life being curious about many aspects of the world around him, particularly the biological world. The extraordinary opportunity of the *Beagle* voyage followed by a life devoid of financial worries produced such enormous dividends in so many ways. It makes one wonder how a similar mind would thrive in today's grant-driven environment. In a way, Darwin had one vital luxury which is in short measure today, and that is time to experiment and think.

This collection of essays would not be complete without an acknowledgement to the many members of Darwin College who facilitated the lecture series, and in particular to Richard and Ann King for their generous financial support, and to Janet Gibson, who brought order to both the contributors and their manuscripts.

William Brown and Andrew Fabian
Darwin College, Cambridge

1 Darwin's intellectual development: biography, history and commemoration

JANET BROWNE

An unusual photograph was published in 1973 that claims to show Charles Darwin's ghost taking a midnight stroll through the British countryside (Figure 1.1). The house and garden are not Darwin's old home, Down House, but a neighbouring mansion in the village of Downe in Kent. According to the photographer, he can be seen in hat and cloak with his walking stick and tell-tale beard (Coxe, 1973).[1] The image poses an intriguing visual puzzle that calls into play our personal belief systems: some of us will be willing to find him, others not.

More to the point, the photograph serves as a metaphor about the biographical process. Historical biographers are in the business of capturing ghosts, creating characters, choosing words and arranging documents to bring people alive again, finding images and seeing shapes, however evanescent, and urging readers to perceive fresh and stimulating pictures in the data presented. Of course biographers work within the given historical record. They do not knowingly present fiction, although Virginia Woolf and other literary critics frequently recommended that biographers should use all the skills of a novelist.[2] Like actors, writers need to interpret and shape their character. The dramatis personae that biographers display in their books are necessarily created figures: people may be presented in a number of ways and made to say a number of things. An important question in the history of science therefore relates to the different ways that scientists have been presented through the ages.[3] One of the most evocative examples must surely be the case of Charles Darwin. A rich variety of images of Darwin have emerged over

Darwin, eds. William Brown and Andrew C. Fabian. Published by Cambridge University Press. © Darwin College 2010.

FIGURE 1.1 A photograph from the 1970s showing Darwin's ghost.
From Antony Hippisley Coxe (1973). *Haunted Britain*. London: Hutchinson. Published
courtesy of the heirs of Antony Hippisley-Coxe.

the years since his death in 1882 – an assortment of ghosts that have
peopled the historical record (Moore, 1996; La Vergata, 1985; Young,
1988; Colp Jr., 1989; Greene, 1993).

Depending on definition, there have been around 30 or so biographical
studies of Darwin since 1882, somewhat more than of Isaac Newton but
fewer than of Marie Curie.[4] The number and variety of these writings
invite extended historical attention (Churchill, 1982).[5] One leading feature
that bears special consideration is the way that the theme of Darwin's
intellectual development has been presented. This is an issue that has
always held cultural relevance, no less diminished in the present day.
For Darwin's fame and intellectual achievement early turned him into
an icon of science, even in his own lifetime,[6] and many writers have
been eager to explore his route to creativity, his education, his developing
self-determination, his reading materials, his inner psychology and those
elements of Victorian culture that may have encouraged the formulation

and expression of his views. The most enduring question that these books ultimately address is how did Darwin become the man who wrote *On the Origin of Species?* How did he change from an enthusiastic amateur naturalist to one of the greatest biological thinkers and authors of the modern era? As Frederick Churchill once argued, the answers to such questions have shifted through the decades according to changes in the way that science has been regarded and in response to diversifying cultural concerns (Churchill, 1982). How writers have dealt with Darwin – and Darwin's intellectual development – indicates something of how people thought more generally about science. The question relates readily to research into other iconic figures in science (Friedman and Donley, 1985; Yeo, 1988; Smith and Wise, 1989; Geison, 1995; Ferry, 1998; Koerner, 2000; more recently, Fara, 2002; White, 2003; Pancaldi, 2003). This article describes a variety of 'Darwins' presented in biographies composed since Darwin's death in 1882 and asks how these might reflect changing concerns about science.

It is helpful first to consider the way in which Darwin himself described the path of his mental development. Darwin was notoriously reticent about his inner life, hardly offering any introspective reflections in the manuscript begun in 1873 that he called 'Recollections of the development of my mind and character'.[7] This document was perhaps not originally intended for publication. Even so, it was published in edited form after his death in Francis Darwin's (1887) *Life and Letters of Charles Darwin.* Much has been made of the way that Darwin's sons Francis and William discussed the matter of publication of this text among the family. Francis Darwin, the self-elected editor of Darwin's *Life and Letters,* and his older brother William, bowed to the wishes of their mother, Emma Darwin, over the content of these personal recollections. At Emma Darwin's request, Francis removed several sentences written by Charles Darwin that expressed contempt about certain points of Christian doctrine (Barlow, 1958). It should be noted that many other passages about Darwin's religious beliefs were left intact: this was not total censorship. Yet Emma Darwin did not want to upset those family members and friends who might be offended by Darwin's frank remarks. It should also be mentioned that for publication Francis Darwin moved parts of his father's text around, dividing it across several chapters in the first volume of *Life and Letters,* re-titling one section as 'Autobiography', and adding other family material as he thought necessary. Darwin's recollections were published in their full and original form only in 1958, an event that forms a later part of this chapter's argument.

In the text available to Victorian readers, Darwin became very real. His recollections were packed with delightful stories about family life, his travels and subsequent career. However, a number of modern scholars have noted how this lively domestic information did not adequately reflect the abundance and reach of his mental life. It seems almost as if he was unable to find in himself anything particularly special or unusual, and that he thought of himself as an amiable and ordinary young man who, by luck and hard work, had turned himself into a natural historian and author. Perhaps this tone of voice is partly explained by the fact that he was writing for his future grandchildren, where family anecdote and reinforcement of diligence and application might have been presumed important. Recent research also suggests that the motif of Samuel Smiles' self-made man was crucial in shaping Darwin's memory of himself (Browne, 2002). Some scholars remark on Darwin's bleak description of the deterioration of his artistic sensibilities, or compare his writing style to that of a taxonomist, holding himself up on a pin to observe and describe. Others look in vain for any moment of epiphany, as might be found in classics of the genre such as Rousseau's *Confessions* or John Henry Newman's *Apologia pro Vita Sua*. Either by chance or design, Darwin left the field open for a great deal of speculation by historians about his intellectual development.[8]

Even so, Darwin there acknowledged his early love for natural history and his passion for collecting natural history specimens, which in later life he came to understand as an urge that might have been satisfied by collecting anything, even postage stamps or biscuits (Barlow, 1958).[9] His delight in collecting mineralogical specimens as a boy in Shrewsbury, his awakening interest in Edinburgh in the larger theoretical explanations underlying natural history and his pleasure in beetle-collecting while at Cambridge University were described by him as key parts of his youth (Figure 1.2). He stated that his time at Edinburgh and Cambridge Universities was wasted as far as the curriculum went. The continuing publication of the *Correspondence of Charles Darwin* can now add a great deal more substance to this account and reveals a much more sophisticated picture of the young undergraduate.[10] Evidently to create a self-portrait

FIGURE 1.2 This caricature of a young Charles Darwin riding a beetle was drawn by fellow beetle collector Albert Way in 1832. It is captioned 'Go it Charlie!' By permission of the Syndics of Cambridge University Library.

of carefree innocence was significant to Darwin. He wanted to tell his story in a particular way. He believed – as indeed we still do today – that the *Beagle* voyage was the turning point of his life. It was the voyage that made him what he became. As he said in these recollections: 'The voyage of the *Beagle* has been by far the most important event in my life and has determined my whole career' (Barlow, 1958).

In this he spoke truly. There were a number of factors he singled out. Darwin spoke warmly of the role that Charles Lyell's book on the *Principles of Geology* (1830–3) played in the development of his scientific ideas during the *Beagle* voyage, providing him not only with information against which to gauge his own observations and theories, but also with an interpretive method that focused on the slow accumulation of many small changes (Figure 1.3). When placed in a chronological framework that stretched over unimaginable eons of time, these small changes could add up to large effects. In Lyell's hands this was a pioneering geological approach. In Darwin's mind, it became a powerful way of thought that could be applied to the whole of the natural world, to biological as well as geological phenomena, and equally also to the human domain. This one idea became the touchstone of Darwin's scientific approach,

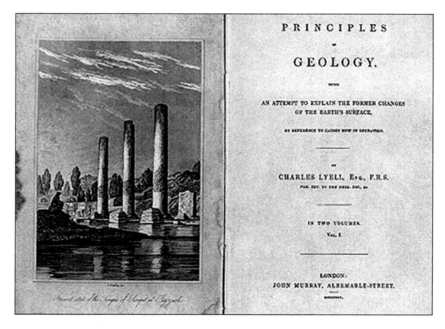

FIGURE 1.3 Darwin took a copy of Charles Lyell's *Principles of Geology* (1830–3) with him on the *Beagle* voyage. By permission of the Syndics of Cambridge University Library.

an intellectual commitment that dominated his work during the voyage and persisted in ever-deepening form through the rest of his career, especially underpinning the central argument of *On the Origin of Species*.[11] There were many other aspects of Lyell's work that were important to him (Herbert, 2005). Later, after his return from the *Beagle* voyage, Darwin valued Lyell's friendship highly. Lyell's impact on Darwin's mind and character indeed can hardly be overstated. He readily acknowledged:

> I have always thought that the great merit of the *Principles* was that it altered the whole tone of one's mind and therefore that when seeing a thing never seen by Lyell one yet saw it partially through his eyes.
>
> (Barlow, 1958)

In these autobiographical recollections Darwin also spoke of the impact of visiting the Galápagos archipelago, a visit that unsettled his assumptions about the fixity of species, and of his adventures in South America, galloping across the pampas in search of animals and plants, or digging

GIGANTIC LAND TORTOISE,—A PRESENT FOR HER MAJESTY.

FIGURE 1.4 A Victorian engraving of a Galápagos tortoise. *Illustrated London News*, 13 July 1850. Author's copy.

for fossils, or seeing the stars from the high passes of the Andes (Figure 1.4). He experienced a major earthquake and (from a distance) saw a volcano erupt, two astonishing examples of geology in action that similarly unsettled his assumptions about the stability of the Earth. He spoke of the experience of seeing indigenous peoples in their native countries, in particular the shock of seeing the local inhabitants of Tierra del Fuego, who to Darwin appeared to live on the edge of barbarity. The shock was deepened by the fact that the *Beagle* carried three members of the Yahmanh tribe who had been captured during the previous *Beagle* voyage and taken to England by Captain Robert FitzRoy for a Christian education (Figure 1.5). The three were being returned by FitzRoy to establish an Anglican mission station. As the *Beagle* men saw it, the three Fuegians on board ship had acquired all the attributes of civilised peoples. Darwin could not help but compare them to their literal relatives on the shore. He was stunned. In the written

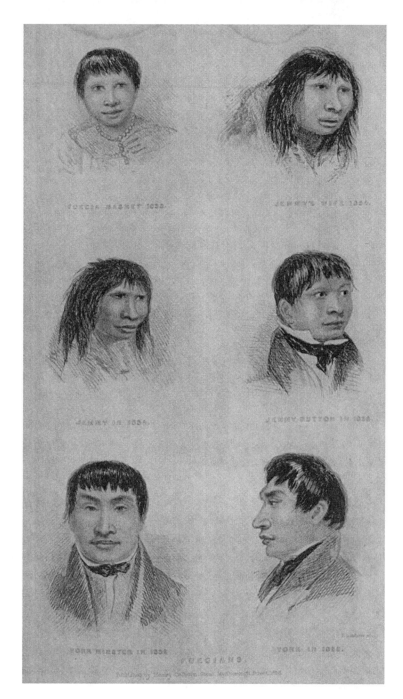

FIGURE 1.5 Three of the four Fuegians whom Robert FitzRoy, captain
of the *Beagle*, captured on an earlier voyage of the *Beagle* and took to England.
FitzRoy intended to resettle them in Tierra del Fuego in a Christian mission station.
From FitzRoy, R. (1839). *Narrative of the Surveying Voyages of His Majesty's Ships Adventure
and Beagle, between the Years 1826 and 1836.* London: Henry Colburn. Wellcome Library,
London.

record he remarked several times on the transient nature of civilisation and the unity that underlay the varieties of mankind (see especially Desmond and Moore, 2009).

These experiences and many others made an indelible impression. Darwin always acknowledged that the *Beagle* voyage was the most formative event in his life. Indeed, it could almost be said that he spent the rest of his life exploring questions that first arose in his mind during those voyaging years, an intellectual journey that took him onwards to the *Origin of Species, The Descent of Man,* and beyond. As Darwin described it in the opening pages of the *Origin*, his *Beagle* experiences had stimulated for him the problem of species. In his autobiography, however, there are no finches, no eureka moment on the Galápagos archipelago.

Instead he said that, on his return from the voyage, he began on the entirely Baconian plan of collecting facts. This too is a fascinating example of the reshaping of memory according to contemporary norms of scientific practice. In Victorian eyes, the name of the seventeenth-century philosopher Francis Bacon was routinely applied to scientific proposals that emerged in an inductive manner from a wide body of facts – the process in itself guaranteed that the proposal was a reliable indicator of a natural truth. In his old age Darwin had many reasons to emphasise his use of an acknowledged scientific methodology. However, his writings from this period tell a different story. To be sure, Darwin did collect many facts about biological change, as evidenced by the now famous notebooks filled by him during the years 1837–9 with a torrent of information and queries about living beings and their possible origin (Barrett *et al.*, 1987), and by the copious correspondence and reading programme that he undertook through the same years and beyond.[12] Current scholarship nevertheless indicates that this information gathering was driven by energetic theorising. Darwin's mind glittered with a wide variety of different possible theories. Nor did he confine himself to the biological world. He read voraciously in the social sciences, anthropology and philosophy. With historical hindsight we can see that this was one of the most remarkable periods of his life – a time of extraordinary intellectual excitement and creativity. To call this 'Baconian' was retrospective lip service to nineteenth-century philosophical convention.[13]

Still, this is the story as Darwin told it. My question is how do the biographers tell it? We can now see that the voyage turned him into a well trained expert. It gave him self-discipline and the tools of his trade, including the ability to make extensive written records. It generated self-confidence, a growing reputation in the eyes of expert naturalists, and provided a network of colleagues and contemporaries who were eager to welcome him into the world of science. His naval connections helped him to distribute his specimens to the major institutions of the day and to plan a broad-based series of publications. More significant for the future of biology, the voyage also opened to him the world of the intellect. Darwin returned to England with a mind bubbling over with fresh ideas. The voyage provided him with countless opportunities to dwell on large, all-embracing questions while learning how to gather precisely focused details. It was truly transformative.

The founding document in the Darwin biographical tradition was the *Life and Letters* (1887), prepared by Darwin's son Francis: this is the same publication in which Darwin's 'autobiography' was first published. This was followed by *More Letters* (1903), edited by Francis and A. C. Seward. Not long afterwards Francis Darwin also produced a single volume on the life of his father, drawing on the material presented in those more comprehensive works. Because these volumes were closely connected to the family, and emphasised family papers and personal correspondence, they were naturally assumed to present an authoritative picture of Darwin. Without wishing to downplay the role of many significant obituaries, especially those that sought to cast Darwin as a hero in science, it seems clear that Francis Darwin's *Life and Letters* served as the primary resource for writers for more than a century after Darwin's death. This enduring printed image of Darwin therefore bears some comment.

Francis Darwin brought personal knowledge, literary ability and scientific understanding to the *Life and Letters of Charles Darwin*, and his volumes, although formulated according to an entirely conventional Victorian genre, were in fact significant in revealing something of the private man. The bulk of the volumes comprised letters, as was to be expected, and these offered behind-the-scenes insight into Darwin's work and preoccupations over a long lifetime. Yet Francis worked hard to present a more substantive account. Introductory pages included the edited version of Darwin's autobiographical recollections, along with a chapter of reminiscences contributed by Darwin's children and friends, and an account by Francis Darwin of his father's working practices and daily life. This last was based on more than just filial piety. For several years Francis had lived at home, acting as his father's secretary and research companion, and he understood a great deal about the meaning of Darwin's botanical interests. Darwin's death – a matter of considerable interest in the era of high Victorian sensibilities – was described briefly in a separate concluding chapter.

These details contributed to the making of a scientific hero (Cubitt and Warren, 2000; Cantor, 1996; Fara, 2000). There were others too. Thomas Henry Huxley, Darwin's most famous friend and public supporter, supplied a chapter on the reception of the *Origin of Species* that cast Darwin as a David to the establishment's Goliath. There were lists of Darwin's publications, portraits and honorary degrees, and an appendix giving the citation for his burial in Westminster Abbey. Mechanically reproduced portrait photographs were reproduced as frontispieces to each of the three volumes (Figures 1.6, 1.7 and 1.8), which became iconic indications of Darwin's presumed persona.[14] Even the arrangement of letters helped manufacture a figure for the era. Francis Darwin judged many letters too technical for a general audience and omitted much about pigeon breeding, for example, or barnacle classification. Letters relating to the *Origin of Species* predominated, indicating that the writing, publication and reception of the book towered over Darwin's other publications. His botanical work was squeezed together into two concluding chapters.

Darwin was throughout described as a loving father and friend, a courteous correspondent, socially respectable, intellectually honest and personally humble – as a man who had assiduously accumulated the mountain of evidence presented in the *Origin of Species*, who stood apart from controversy and nobly endured lifelong illness. As the son painted it, these fine, simple qualities were crucial factors in Darwin's intellectual success. He wrote, 'In choosing letters for publication, I have been largely guided by the wish to illustrate my father's personal character' (Darwin, 1887, p. iii).

FROM A PHOTOGRAPH (1854?) BY MESSRS. MAULL AND FOX. ENGRAVED FOR
'HARPER'S MAGAZINE,' OCTOBER 1884.

Frontispiece, Vol. 1.

FIGURE 1.6 Charles Darwin. Wood engraving by G. Kruell from
a photograph by Maull & Fox, London, circa 1854. Frontispiece to Darwin, F. (1887).
Vol. 1. Author's copy.

Nowadays we can understand much of this emphasis on humility
and respectability as the Victorian equivalent of the credit-building
processes described by Steven Shapin in seventeenth-century science
(Shapin, 1994; see also Shapin, 1991). It was important for Victorians

FROM A PHOTOGRAPH (1874?) BY CAPTAIN L. DARWIN, R.E. ENGRAVED FOR THE
'CENTURY MAGAZINE,' JANUARY 1883.

Frontispiece, Vol. II.

FIGURE 1.7 Charles Darwin, seated, Down House, Kent. Wood
engraving by T. Johnson, 1883, from a photograph by Leonard Darwin, circa 1874.
Frontispiece to Darwin, F. (1887). Vol. 2. Author's copy.

FIGURE 1.8 Charles Darwin, standing, Down House, Kent, 1881.
Wood engraving from a photograph by Elliott & Fry, London, circa 1880.
Frontispiece to Darwin, F. (1887). Vol. 3. Author's copy.

to believe that science was carried out by trustworthy individuals
who did not seek to overthrow either the church or the political establishment. No words can be strong enough, said the American biographer
George Woodberry in 1890, to express the moral beauty of Darwin's
character. He was a marvellously patient and successful revolutioniser
of thought, declared George Bettany in his 1887 biography. Enough of
his character shone forth in his work to indicate his tenderness and
goodness, added Archibald Geikie in 1888. A life of singular purity,
thought Charles Frederick Holder in 1891. 'Translucent truthfulness'

said Henry Fairfield Osborn, of the American Museum of Natural History in 1924. Darwin's virtuous character was evidently being put forward by these first biographers as an important element in his moral right to speak about the natural world, a key element in science's claim to uncover truths in Nature, no matter how controversial. Darwin's personal humility and goodness were here construed as part of the process of convincing contemporaries that discussions of evolutionary theory could be conducted in a rational, courteous manner and were not dangerously atheistic or subversive.

Most of all, however, Francis Darwin portrayed his father as a hard worker. To showcase this mental labour was another significant biographical move that took its impetus from the busy, productive, self-made man of nineteenth-century industrial society as described by Samuel Smiles.[15] Earlier ideas of the uniquely inspired individual promoted by Thomas Carlyle and Ralph Waldo Emerson, or the romantic hero embodied by Humphry Davy, Goethe, Keats or Lord Byron, were almost completely supplanted in certain sectors of British society in the last third of the century by Smiles' eulogies of honourable toil – a shift away from the terminology of 'genius' towards 'exertion', from inspiration to perspiration. These notions of industriousness contributed to new ways of representing science emerging in Europe and North America at the end of the century. The Christian topoi of saintly self-dedication and austerity were transferred to the idealised man of science. Great thinkers were rewritten as great workers. Looking back over history, said Smiles, it can be seen that Louis Pasteur was notable for his extraordinary scientific perseverance. Tycho Brahe scarcely left his observatory for 21 years. William and Caroline Herschel led patient and laborious lives. Even Pliny never relaxed, except in his bath. Celebrated figures, repeated Smiles, were careful economists of time (Smiles, 1887). In essence, Darwin was presented in Francis Darwin's writings, and thence in other Victorian biographies, as an industrious man for the industrial age. He was revered as an independent and assiduous thinker, dedicated to his work, someone who nobly struggled against illness. It was a life in which individual effort, achievement and personal virtue in the face of difficulties were heavily featured.

A very different Darwin appeared in the aftermath of World War I. Hard work and virtue gave way to an emphasis on inner turmoil. The 1920s and 1930s saw those authors who wrote in English turn towards the interior life of men and women, answering a new social and psycho-analytic call to acknowledge the many-sided self and a search for greater authenticity in representation. Biographers experimented with modern-ism and felt that their special focus on human character contributed significantly to the literary restructuring of the period. It was the day of the biographer, wrote Hesketh Pearson in 1930, a prolific biographer himself whose first publication was a study of Dr Erasmus Darwin. There was much theoretical consideration of life-writing, taking its colour from Lytton Strachey, André Maurois, Emil Ludwig, Leslie Stephen and Edmund Gosse, appearing most obviously in Britain in the writings of Virginia Woolf and Harold Nicolson, and in America in those of Gamaliel Bradford, who advocated a new genre of 'psychography'. Bradford was not overly interested in the content of Darwin's science except insofar as this made him one of the makers of the modern world. Yet he wrote an excellent short biography of Darwin in 1926, one of the first to ask the modern question, what kind of man was this, so sick and so secular.

Creativity featured as an issue too. Nora Barlow, Darwin's grand-daughter, one of the new breed of scientific women actively researching Mendelian genetics with William Bateson in Cambridge, began in the 1930s to explore the origins of Darwin's evolutionary ideas, seeking to understand his inspiration and motive. Though she did not write a biography as such, she was a pioneer in bringing Darwin's unpublished manuscripts to bear on the question of his intellectual development. Nora Barlow holds an esteemed place in Darwin scholarship and is especially noted for her 1933 edition of Darwin's *Beagle* diary. This document gave a day-by-day immediate picture of the events of the voyage as experienced by Darwin and, for the first time, opened up the *Beagle* years to new forms of scholarship. This diary was the source for the cleaned-up, published account given in Darwin's *Journal of Researches* (1839), commonly known as *Voyage of the Beagle*. Barlow followed through in 1945 with an edition of letters and notebooks from the *Beagle* voyage (Barlow, 1933; followed by Barlow, 1945). She later published the full version of Darwin's autobiography with omissions restored, and

Darwin's 'Ornithological Notes', written at sea on the way back to Britain, in which he first posed the problem of species (Barlow, 1963; see also Barlow, 1967). Her commitment to disclosing the inner Darwin through unpublished manuscript sources such as letters, autobiographical recollections, notes and diaries marked a turning point in Darwin studies. Though we may now think such an interest in the private records of a great thinker is only to be expected, this focus on manuscripts was relatively new in the history of science and reflected a wider fascination with the course of psychological development and the personal lives of others. A material factor in this realignment in Darwin studies was the magnificent bequest, starting in 1942, of papers owned by the Darwin family to Cambridge University Library, with the support of the Pilgrim Trust (Burkhardt and Smith, 1985).

Importantly, Bradford and Barlow both considered Darwin as one of the secular moderns. To them the fading of doctrinaire religious opinion was the key to the rise of modern science and they set about publicising Darwin's loss of faith in order to promote what they saw as the new dawn of rational thought. Neither were militant atheists. Barlow in particular saw Darwin's good personal relations with religious individuals such as his wife as a significant matter. In a series of appendices to her edition of Darwin's *Autobiography*, she published important family documents, among them letters from Emma Darwin to her husband about his declining faith (Barlow, 1958, appendices). These indicated that although Darwin's approach might be entirely naturalistic, his moral world was eminently respectable and that he thought carefully and considerately about the implications of his theories. Louis Trenchard More's 1936 biography of Newton similarly made use of the Hurstbourne archive to reveal the full extent of Newton's interests in unorthodox theology (Higgitt, 2007). These post-war thinkers were perhaps struggling with the bold dichotomies expressed by Lytton Strachey's *Eminent Victorians* in 1918. Strachey's original idea was to write 12 biographical silhouettes, including Ellen Terry, the Duke of Devonshire, Charles Darwin, Benjamin Jowett and John Stuart Mill. But Strachey felt he could not include Darwin, who he regarded as being on the side of the moderns. Had his initial intention been fulfilled, the work could not have had the same debunking effect (Holroyd, 1994).

As part and parcel of this new biographical focus, Darwin's house (Down House in Kent) was purchased and opened as a museum in 1929, a temple to his mind and personality as they were then perceived (Figure 1.9). The museum was as biographical in content as any text between two covers. It was furnished with the help of surviving family members who recreated the old rooms and gardens as they remembered them.[16]

FIGURE 1.9 Down House, Kent, in the years after Darwin's death. Wellcome Library, London.

FIGURE 1.10 Darwin's study at Down House. From a photograph taken after the opening of the house as a museum in 1929. Wellcome Library, London.

The acquisition was intimately tied up with scientific concerns that paralleled the literary shifts just described.[17] As an expression of Darwin's personality, the Down House museum portrayed British science at its most ingenious and reassuring. The traditional gardens, country setting and old-fashioned furnishings indicated that Darwin's science had been essentially domestic, humane and far removed from a laboratory setting. All that Darwin had needed for his great work, visitors might think, were his eyes, and a pencil and paper. Darwin's original study, where the *Origin of Species* was written, was encountered as a secluded, homely place, markedly different from the fast-moving research institutes active elsewhere, even including a stuffed dog, ordered from a taxidermist, curled up in a basket beside the fire (Figure 1.10). Such a home implied that scientific knowledge posed no threat to cultural values. While the effects of the Scopes trial of 1925 cascaded through the United States of America, Darwinism in Britain was presented as a domestic entity, very far from undermining religious belief. Instead, the museum showed Darwin as a clear-eyed observer of Nature, a man whose personality and conservative habits of life indicated that any dangerous thoughts were in safe hands.

The next flurry of biographies came with the 1959 centenary celebrations of publication of the *Origin of Species*. As in most commemorative events there was great interest in recalling and reconstructing the life of the individual who was being celebrated (Abir-Am and Elliott, 1999; Jordanova, 2000a; James, 2008; Browne, 2003). The ground was fertile for some writers to promote another kind of Darwin, one who was more obviously biological in focus. By then, the self-labelled evolutionary 'synthesis', as worked out in previous decades by Ernst Mayr, Ledyard Stebbins, George Gaylord Simpson, Julian Huxley and influential others, had taken shape (Mayr and Provine, 1980). As Betty Smocovitis argues, the Darwin celebratory year of 1959 was perceived as an appropriate moment to affirm this 'synthesis' and also make direct links back to Darwin as a magisterial ancestral figure who put the theory of evolution by natural selection on the map (Smocovitis, 1999). There were interesting elements of self-validation at work here, neatly illuminated by Smocovitis. Several of the active participants in the modern synthesis wrote biographies of Darwin or explored his work with reference to the scientific developments of the day. These authors, newly confident that evolutionary theory was true, went out of their way to remind readers that Darwin's ideas held an active place in current thought. Ernst Mayr, for example, who proposed the 'biological species concept', was immensely significant in reappraising Darwin in the light of modern biology, building in the process his own reputation as a biologist–philosopher–historian. An old-style European positivist, Mayr encouraged biologists to regard Darwin as the man who was the source of all the best modern biological ideas. Several scholars note Mayr's tendency to put his own work at the heart of the new Darwinism, perhaps even thinking of himself as a Darwin for the modern era, and certainly in books such as *The Growth of Biological Thought: Diversity, Evolution, and Inheritance* (1982) and *One Long Argument: Charles Darwin and the Genesis of Modern Evolutionary Thought* (1991), where he restructured the history of biology to suit his own purposes.[18]

At the same time, skilful writers for the public like Howard Mellersh and Alan Moorehead published accessible studies about Darwin's observations on the *Beagle* voyage (Mellersh, 1964; Moorehead, 1969). These too celebrated Darwin's observational skills, his biological insights and the role of the *Beagle* voyage in the development of his later views. As an aside, it can be noted that the *Beagle*, rather than the *Origin of Species*, took pride of place in these 1959 celebrations, again confirming in the reader's mind the transformative role of the voyage in Darwin's life. Another well received study was published in 1955 by Arthur Keith, *Darwin Revalued*. Keith was a notable comparative anatomist and expert on the human fossil record, and also a key advocate for the preservation of Down House. His book, like Mayr's, displayed a deep appreciation of Darwin as a practising scientist. In about 1930, Keith went to live in Downe village, acquiring a special empathy for the historic surroundings: 'Darwin became a real man for me: I saw him as he moved about from day to day'.[19]

With this 'biologised' Darwin, the Galápagos finches came into their own. First described by David Lack as an exemplary case of evolutionary radiation in 1947, Lack's book was reissued with a new preface in 1961. These materials supported the view that Darwin's theory was, in its essentials, based on biological observations made on the Galápagos Islands (Figure 1.11). Frank Sulloway explains that Lack's writings became the source for a widespread belief that Darwin had experienced a 'eureka' moment in the archipelago, to which should be added Nora Barlow's edition of 'Darwin's Ornithological Notes' in 1963 (Sulloway, 1982; Barlow, 1963). The possibility of such a discovery moment fitted well with contemporary images of how science advanced.

It also tied in with plans first circulating in the 1950s for a Darwin memorial park on the Galápagos Islands. Julian Huxley, the first Director of UNESCO, founding member of the World Wildlife Fund, and the grandchild of Thomas Henry Huxley, was instrumental in initiating the process of designating the islands as a World Heritage site (ultimately achieved in 1978). Huxley delivered several lectures in the 1959 anniversary period that publicly associated Darwin with secular science. In *The Living Thoughts of Darwin* (1937), reissued in 1959, he described Darwin as a rational, scientific figure, 'the Newton of biology' (Huxley, 1959). And at a commemorative meeting of biologists and philosophers in Chicago, he promoted the view that religion was merely

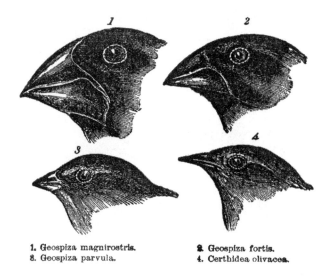

1. Geospiza magnirostris.   2. Geospiza fortis.
3. Geospiza parvula.   4. Certhidea olivacea.

FIGURE 1.11 The beaks of four Galápagos finches showing modifications in shape according to foodstuffs. From Darwin, C. (1845). *Journal of Researches into the Natural History and Geology of the Countries Visited During the Voyage of H.M.S. 'Beagle' Round the World.* London: John Murray. Wellcome Library, London.

a Darwinian adaptation in humankind (Smocovitis, 1999, pp. 302–5). Indeed, Julian Huxley seems to have served in those years as a visible and literal descendant of the controversies of the 1860s, in which Thomas Henry Huxley had jousted with Samuel Wilberforce at the Oxford meeting of the British Association for the Advancement of Science (James, 2005).[20] Along with Ernst Mayr, the younger Huxley received the Darwin–Wallace medal of the Linnean Society of London in 1958. He was knighted in the same year, 100 years after Charles Darwin and Alfred Russel Wallace jointly announced the theory of evolution by natural selection. Honours like these marked the new evolutionary biology as a leading scientific discipline.

In similar vein, Gavin De Beer, director of the British Museum (Natural History) in London, saw the centenary year as a fine opportunity to celebrate Darwinism and to bring forward new editions of Darwin's private writings. In 1958 De Beer arranged the XVth International Congress of Zoology as a commemorative tribute to the reading of the Darwin–Wallace paper at the Linnean Society of London. He then published an edition of Darwin's evolutionary notebooks. Publication of these extraordinary notebooks opened a new phase in Darwin studies, although De Beer was hampered by the necessity of working for some time with incomplete documents (De Beer, 1960; Churchill, 1982). Several pages had been excised by Darwin when he prepared what was to become the *Origin of Species*.[21] The excised pages were not located for many decades after Darwin's death, and only reappeared in the 1970s (Burkhardt and Smith, 1985). On one of those pages Darwin recorded in September 1838 his reading of T. R. Malthus' *Essay on the Principle of Population* (1798) and made an explicit first statement of the idea of natural selection derived from Malthus' account of the struggle for survival.[22] In his 1965 biography of Darwin, De Beer naturally took special interest in the early development of Darwin's theories as explored in the notebooks (De Beer, 1958; 1965). Yet De Beer was proved wrong when he downplayed the role of Malthus in Darwin's path to formulating the concept of natural selection. For De Beer, the proper emphasis lay instead in Darwin's talent for biological observation.

Many biographies written around that time presented Darwin almost as if he were a living biologist encountering the same range of scientific issues as the men and women of the 1950s. Non-biological writers had no such vested interests. Gertrude Himmelfarb and Loren Eiseley refused to accept any divide between science and its social and political context, and argued that Darwin should be understood as an individual embedded in Victorian society. Both held a bleak view of the liberal west in the 1950s. Eiseley thought that the modern scientific enterprise had pushed humanity too far away from its sense of responsibility to the natural world (Pitts, 1995). In *Darwin's Century* (1958), he placed Darwin in the context of other evolutionary thinkers which – in a centenary volume – somewhat reduced his heroic role. Later, in *Darwin and the Mysterious Mr. X: New Light on the Evolutionists* (1979), Eiseley suggested that Darwin appropriated the concept of natural selection from Edward Blyth, an early correspondent, thereby undermining what was then believed about Darwin's personal moral code (Eiseley, 1958; 1979). Eiseley's practical knowledge of biological science and his abiding sense of the mysteries inherent in the natural world thus led him to regard evolutionary materialism with little favour. Wallace biographers began to take something of the same critical approach toward Darwin in the 1980s (Brackman, 1980; Brooks, 1984; continued by Shermer, 2002; Slotten, 2004). We can see here the earlier image of a heroic Darwin breaking down into that of an ordinary mortal who experienced ethical dilemmas and pressures, just like anyone else. Science was beginning to show feet of clay.

Himmelfarb's biography of Darwin (1959) also generated considerable interest. Noted for her neo-conservative attention to issues of virtue, morality and the promotion of Victorian values, Himmelfarb based her interpretation squarely on the implications of evolutionary theory for the human race, especially dwelling on Darwin's use of Malthus' political economy. One of her other books was an extended study of Malthus and the need for continuing recognition of Malthusian principles in modern life. She saw social determinism in Darwin's writings, arguing that he failed to take moral responsibility for the theories he put forward when they were applied to humans. This study of Darwin's life gave Himmelfarb an opportunity to argue that moral principles were as much a part of public discourse as of private discourse. She was criticised by biologists for dethroning an icon but also censored by the developing field of history of science for being too keen to moralise.

Again, the image of a saintly Darwin was weakening. Around this time, Saul Adler categorised Darwin's illness as mere hypochondria, providing a modern psychoanalytic interpretation of what the Victorians had regarded as heroic suffering. Donald Fleming diluted Darwin's creative imaginative life by describing him as an 'anaesthetic man'.[23] Historians of science, too, were beginning to turn away from the valorisation of heroes, the 'great man' school of thought, and were instead busy shrinking their subjects into real people. New standards were being created by a growing number of professional historians of science and biography was increasingly regarded as inappropriate for a scholarly discipline anxious to consolidate its academic credentials. With some notable exceptions, it has taken nearly 40 or 50 years for the genre to be re-evaluated as a suitable – if only occasional – vehicle for larger theoretical and interpretative proposals.

Today, biographers work hard to reconstruct the context in which Darwin lived, the problems he tried to solve, the resources upon which he drew for answers and the ways that readers responded to his writings.

This has meant dismantling the imagery of Darwin as an isolated thinker who independently revolutionised the Victorian world view. Instead, Darwin is depicted as an active member of concentric circles of practitioners, an intensely social being, always gauging and fine-tuning his arguments, eagerly gathering support in order to present his theories effectively. Modern writers are also highly attuned to the other evolutionary writers of the period and are deeply conscious of the problematic nature of claiming a solely 'Darwinian' revolution, or even claiming a revolution at all.

Above all, these studies have been transformed by access to a wealth of manuscripts, printed materials and other documents. Peter Brent (1981) and John Bowlby (1990) were among the first to use these rich resources for subtle and informative analyses. As the importance of new historiographies became established, and Darwin's correspondence became more widely known through the 1980s, scholars also began to pioneer vigorous sociological interpretations. Desmond and Moore's deservedly celebrated biography (1991) displays Darwin fully embedded in social and cultural context. My own biography presents Darwin as a networker, a man at the centre of overlapping circles of friends, neighbours, correspondents, enemies and defenders. In the celebratory year of 2009, a number of other studies were produced. All are interested in the making and circulation of knowledge and the way in which new proposals were handled by the community. To write a scientific life nowadays is to provide a framework on which larger social and historical edifices can be built.

What can be said today about Darwin's intellectual development? Modern studies certainly indicate that Darwin came to construct his theories over a longer period of time and with the help of many more people than was previously thought – that his moments of insight depended not only on the observations he made *in situ* during the *Beagle* years, although these must always take their place in any assessment of his mental repositioning, but also on the input of other thinkers, such as John Gould (who identified Darwin's Galápagos birds) or the many acquaintances and correspondents who discussed or facilitated his findings. His theory of evolution by natural selection was not just based on the writings of T. R. Malthus but also on material derived consciously and unconsciously from a wide range of contemporaries. Darwin's mature work, for example, could not have emerged in the form that it did without inspiration from Charles Lyell. The *Beagle* expedition's success depended on Robert FitzRoy. Darwin's invitation to travel came through the good wishes of Professor John Stevens Henslow, who then advised and supported him as a mentor through the travelling years and beyond. Moreover, some of Darwin's earliest ideas about species, deriving from conversations with country gentlemen, were retained and incorporated into the theory of natural selection after he read Malthus in 1838, such as an abiding belief in an analogy between the practices of animal and plant breeders and the processes of Nature. His own work in assessing and publishing the voyage's scientific results contributed too – not least in moving him to ask what was the purpose of the amazing variety of living Nature that he had experienced as he travelled around the world. We also know from the rich documentary record that Darwin's developing views about the place of humans in the natural world, his self-examination of his personal religious beliefs and his relationship with his wife and other close friends encouraged him to explore grand philosophical questions about the human condition with sensitivity and courage. So while it will never be possible exactly to pinpoint Darwin's trajectory of creativity, we have remarkable opportunities for generating new insight into the process.

More generally, it has become clear that the *Beagle* voyage and the notebook years were only two stages in Darwin's intellectual journey. The new sociological perspectives in the history of science encourage a much more broadly based vision. Nowadays, the voyage is recognised as a geographical, political enterprise that served to enhance British affairs overseas, and Darwin can be seen as an active participant in that venture, a player in a larger story about the rise of the British Empire and the globalisation of science. He was lucky to be able to take advantage of these historical shifts – shifts that provided him with a niche in the emerging scientific world as an acknowledged expert on natural history. His family background was sufficiently wealthy to have permitted his voyage around the world as the captain's guest, and to live thereafter in London and Kent as a gentleman with a private income: Darwin never needed a job, such as Thomas Henry Huxley or Joseph Dalton Hooker had, and this freedom from professional employment gave him intellectual independence as well as time to write and ponder. These social advantages gave Darwin a position of respectability that moderated some of the controversy when he published *Origin of Species*. And careful research into Darwin's university training indicates that he was a well educated and talented young naturalist, far from the ingénue that he liked to characterise. He was suitably qualified to accompany a voyage for natural history purposes. This was tacitly acknowledged by Darwin's Cambridge professors, especially John Stevens Henslow, but also shown by the manner in which the invitation to join the *Beagle* voyage came through a series of social networks involving the great men of the university and British Admiralty. In this sense, Cambridge University gave Darwin his future.

The biggest impact on established ideas has been the publication of Darwin's voluminous correspondence in a continuing edition by Cambridge University Press from 1985. This edition reveals that much of Darwin's greatness lay in his dedication to exploring every last ramification of the theory of evolution by natural selection. One major consequence of the correspondence now available is that it becomes clear that the creative moments in Darwin's life continued far beyond the *Beagle* period. Darwin relentlessly pursued his ideas through the 1840s and 1850s, always alert to the possibility of changing his mind, seeking answers to the difficulties that readers would encounter in his views, constantly expanding and exploring, and taking on new investigations such as the classification of barnacles, the instincts of social insects or the breeding of pigeons to substantiate his points. His letters demonstrate that he was much more of an experimentalist than previously suspected. They also demonstrate that he extended his theories substantively in the 1850s, for example to include the major new idea of divergence.[24] The realisation that Darwin constantly adjusted his ideas in sophisticated and significant ways in the period immediately before writing the *Origin of Species* transforms our understanding of the historical record.

This process of reassessment and research was not just confined to the days before the *Origin* either. The letters tell us that Darwin continued to extend his ideas after publication through a succession of important texts, including *The Descent of Man* (1871) and *Expression of Emotions* (1872), and also in the botanical writings that served as explicit examples of evolution in action yet are customarily underestimated by scholars.

Moreover, the existence of this vast archive of letters has opened up new ways of thinking about the manner in which Darwin – and perhaps others – actually conducted their researches. Darwin systematically used the nineteenth-century postal system to facilitate his work (Figure 1.12). At every stage of investigation, and no matter how abstruse a topic, he solicited information by letter from an increasingly large network of correspondents, progressively reaching across the globe to Australia, South Africa, Brazil and wherever else he could find a knowledgeable contact. Sometimes his correspondents were already known to him through their publications or personal connections. Other times he sought introductions to those experts who could best supply the information he needed. He also found informants across very diverse groups of people, counting clergymen, anthropologists, physicians, country gentlemen and gamekeepers among his regular correspondents, and valuing contacts with a number of women and many family members with practical experience of particular issues. This fresh view of Darwin's working practices reveals a different Darwin from the one formerly presented in biographies. It shows him as an efficient networker and gatherer of facts. We see how he contributed to the controversies over his writings – never taking the front of the stage as friends like Huxley were willing to do, but actively participating by letter from his study in Down House.

Here, we are coming to understand Darwin as a man rooted in his cultural context, constantly improving his ideas, vigorously participating in the evolutionary debate, engaging over many decades with the new fields of research that were emerging and discussing matters of the day with an extremely wide range of correspondents.[25] In a general sense, then, moments of creative epiphany are not as central as they once were to our understanding of a scientist's development. Scholars nowadays place far more emphasis on the working practices of great figures like Darwin and on the social conditions that made him visible – themes that hardly appeared in biographical writings before the 1970s. To some extent modern historians are dealing less with the details of

FIGURE 1.12 Much of Darwin's research depended on his correspondence. This image of a Victorian postman derives from the popular song of the day 'The Postman's Knock'. Mary Evans Picture Library.

the science and more with the cultural features that create a scientist and validate a new or controversial theory.

In conclusion, it seems that significant aspects of Darwin's intellectual development have regularly been reformulated by biographers. These authors chose to present slightly different pictures of Darwin, each one attuned to the science of their period, moving from a hardworking, industrious Darwin in the late nineteenth century, through a respectable, domestic man in the 1930s and the biologists' Darwin of the 1950s, to a dedicated networker and letter writer in the 1990s. Perhaps this is not overly surprising. I would not wish to claim that these various images were mutually exclusive, because elements of each were readily available in the literature and in many instances can complement each other. Yet it is important to acknowledge that the people who wrote about Darwin at specific times in the nineteenth and twentieth centuries were remembering the past partly in order to think about their present. They inhabited a different memory world from each other and from us. Science itself is under examination in these books. The way in which Darwin's intellectual development has been discussed in a long and notable procession of biographies provides considerable insight into those different worlds.

2 Global Darwin

JAMES A. SECORD

The most evocative room in a writer's house is almost always their place
of work. Such settings are carefully arranged to create a sense that the
famous occupant has only just left the room, whether it be the print-lined
study of the naturalist Carl Linnaeus at Hammarby in the Swedish
countryside or the novelist Victor Hugo's room on the upper floor of
Hauteville House in St Peter Port, Guernsey, with its linen canopy used
to dry the pages of the works that flowed from his pen. The most
memorable images of Charles Darwin that appeared after his death in
1882 were not portraits, but photographs and etchings of his study at
Down House (Figure 2.1). This was where Darwin wrote *On the Origin of
Species, The Descent of Man* and most of his other books. Engravings of
the botanist Joseph Hooker, the geologist Charles Lyell, and his grandfather Josiah Wedgwood hang above the mantelpiece; books, visible in
the mirror, line the opposite wall; horizontal shelves for filing notes are
on the right. Pilgrims today visit it as a shrine, wending their way
through the London suburbs into rural Kent to see about as sequestered
a place as exists in the history of science. Millions more have seen
Darwin's study lovingly recreated as a room set in the travelling exhib-
ition organised by the American Museum of Natural History in
New York. It is a classic instance of the *genius loci*, the spirit of place.

Look closely, however, and you can see that the desk is littered with
letters, documents, journals and articles from around the world. These
evoke a different set of associations. Far from an isolated backwater,
Darwin's study was a key intersection in a network of intellectual

Darwin, eds. William Brown and Andrew C. Fabian. Published by Cambridge University
Press. © Darwin College 2010.

FIGURE 2.1 Darwin's old study at Down House. CUL, Darwin archive.
Reproduced by permission of the Syndics of Cambridge University Library.

intelligence and trade, connected through the nineteenth century's first
and most significant global communications system: the post office. His
correspondence, which is being published both on the internet and in
some 30 printed volumes, is vast: over 15 000 letters to and from him,
and more are being discovered all the time (Burkhardt *et al.*, 1985–).
These bring out better than anything else Darwin's ways of working.
They record a host of everyday details, from the coming and going of
servants to the arrival of specimens and accounts of new discoveries. On
a typical day Darwin might receive letters from a missionary in Africa,
an informant in a botanical garden in Calcutta, and a doctor in the
American Midwest. He might write to a young professor in Beirut or a
zoological collector in the Malay Peninsula. The great virtue of the
correspondence is to remind us that even the most intimate aspects of
intellectual work were embedded in networks of cross-cultural exchange.

The aim of this essay is to look outwards, to show how the writings and reputation of a single author became central to debates about the role of science throughout the world, and to ask what is 'Darwinism' and how did it become significant in so many places?

The transformation of science

Darwin entered the sciences at one of the most exciting times in the history of science. Not only were individuals developing new theories; more fundamentally, the very idea of what it meant to be a scientist and how to do science were forged. The foundations of knowledge, where to go, and what to do there, were hotly debated. Distinctions between ways of knowing that had persisted since the ancient Greeks were broken down. On the one hand, the tradition of natural history had involved collecting, inventory and display. Although transformed during the Renaissance by the exploration of the New World, and again during the era of explorations in the eighteenth century, natural history retained considerable continuity. On the other hand, the tradition of natural philosophy had aimed to understand the origins of new forms. By 1800 these divisions began to collapse as analysis – penetrating beneath the surfaces of the thing being described, towards an understanding of the causes of diversity – became a dominant mode of enquiry. Form, structure and function were integrated in a much closer way. As a result, new sciences came into being, from 'physics' to 'comparative anatomy', and especially 'geology'. Those practitioners able to combine a range of disciplines under the new analytical order and apply them on a global scale were the most admired: the traveller Alexander von Humboldt, the astronomer John Herschel, and the philosophical geologist Charles Lyell. So Darwin entered science at a moment of great ferment, when becoming a naturalist involved embracing fundamental changes in the shape and order of knowledge, and in viewing the Earth as a whole.

It was this transformation that led the British Admiralty to approve the *Beagle*'s circumnavigation of the globe, on the voyage that Darwin acknowledged as the most important event in his life. Popular myth to the contrary, the voyage wasn't designed especially to take Darwin to the Galapagos in order to discover natural selection. Instead, the *Beagle* was sent to chart the southern coasts of South America, as the newly independent countries in that region had recently opened their ports to British trade. In addition, Captain Robert FitzRoy had orders to make a chronometrical survey of the globe using very accurate instrumentation: this would aid in determining longitude for trading vessels. Darwin was a paying passenger on an imperial voyage. Nearly all the scientific problems that Darwin solved on the voyage, such as the origin of coral reefs, emerged from this programme of global research at the heart of the new sciences.

From the start an understanding of humans was part of Darwin's enquiry. The voyage provided a panoramic vision of human diversity, from Africans on mid-Atlantic islands to the Malay people of the Indian Ocean. Even on the Galapagos, Darwin was at least as impressed by the varieties of the different people (and especially their moral character) as by the celebrated finches. His most shocking encounter was with 'savages' in the windswept wilds of Tierra del Fuego – people imagined on the lowest rung of a hierarchy of races, with apparently rudimentary speech and the most primitive forms of organisation. Yet in stressing their utterly alien, 'hellish' existence, Darwin immediately came back to questions of origins and inheritance. These men were brothers: 'such were our ancestors' (Darwin, 1871).

Back in London, living in a dingy bachelor flat in the bustling centre of the largest city the world had ever seen, Darwin formulated his principle of natural selection after reading the Reverend Thomas Malthus' *Essay on the Principle of Population* (1826, 6th edn). Malthus had discussed human populations throughout the world, showing that scarce resources led to a struggle for survival. All but a few would die before being able to reproduce their kind, with the weakest individuals losing the battle for scarce resources. Writing in his notebook, Darwin compared this struggle to the force of a hundred thousand wedges smashing into a single point on the face of Nature. This was not the homely, barnyard perspective of his later analogy with domestic selection as practised by cattle breeders and pigeon fanciers; it was Nature seen with an abstracted view from outside and above, what Herschel had called 'the eye of reason' (Herschel, 1830). At the very moment of his great discovery, Darwin relied on the distanced, analytical perspective made possible by the global ambitions of the new sciences.

The all-encompassing vision at the heart of the questions Darwin pursued on the voyage remained with him for the rest of his life. Through the efficiency of the postal system, Darwin's desk became an instrument for engaging the world, at least as significant in this respect as the more celebrated *Beagle*. Besides letters, the post brought many other forms of information to Darwin's attention: he read the *Athenaeum*, with reports of the London learned societies; heavyweight quarterlies such as the *Westminster* and *Edinburgh*; the *Gardener's Chronicle*, with its weekly tidbits of horticultural and botanical news; and scientific journals such as *Nature* and the proceedings of the societies of which he was a fellow. The family read *Punch*, the *Illustrated London News*, and the daily news in *The Times*. Publishers forwarded copies of reviews; authors from Boston, Sydney and Paris eager to have their work read sent books, offprints and clippings. Wheeling around his study on a specially designed chair, Darwin could vicariously travel the globe, secure in what Janet Browne has rightly called his 'ship on the Downs' (Browne, 1995). As he later said, he was a 'complete millionaire in odd & curious little facts';[1] as such, he represented the apotheosis of the gentlemanly capitalist upon which the British Empire was founded, a characteristic figure of the first great age of globalisation. Engagement with the world beyond the parish, the county, the nation and Europe thus affected Darwin, not only through his personal experience as a traveller and a cosmopolitan gentleman, but through the ways in which the making of knowledge was transformed during the early nineteenth century.

A history of reception

If it is now well known that Darwin's work depended on imperial networks and global perspectives, investigating the emergence of Darwinism throughout the world inevitably is a more challenging task. This wider story is much less well understood, despite the reception of Darwin's writings and the spread of his reputation having been staples of historical enquiry for over half a century. The first serious study was written in 1958 by the Swedish scholar Alvar Ellegård, who used statistics in a form typical of sociological study in the 1950s, characterising readers as 'lowbrow', 'middlebrow' and 'highbrow' (Ellegård, 1958).

Ellegård confined his attention to Britain, but from the 1970s studies of Darwin's reception have widened to include many other countries, from China and Russia to Latvia and Australia. The results have been fascinating, but an almost unremitting focus on the nation state as the unit of analysis has stymied their usefulness. Reading many of these studies in succession is rather like attending a late nineteenth-century world's fair, with a distinct pavilion for each country. After a certain point the effect is numbingly repetitive: usually there is a group of pro-Darwinian materialists, various manifestations of traditionalist opposition, and then, gradually, a process of accommodation and compromise.

The problem is not limited to the specifics of Darwin and Darwinism. Despite a huge interest in various forms of world history during the past two decades, until recently intellectual history and the history of science have lagged behind in thinking beyond national boundaries, particularly those of the United States and Western Europe. All too often, the failure of attempts to draw together a larger international picture is the result of the lingering legacy of the history of ideas, in which intellectual controversies are dissected primarily on conceptual lines. Thus 'Darwinism' is first separated into what Arthur Lovejoy called its 'unit ideas' ('evolution', 'natural selection', 'man' and so forth) and the fortune of each is traced. The components vary – the great naturalist Ernst Mayr identified Darwin's 'five theories' (Mayr, 1985), leaving out human evolution as not conceptually distinct from a biological point of view – but the result is inevitably a story of diffusion, dilution and distortion, as Darwin's ideas move further and further from their originating point in the study at Down.

If such surveys remain unsatisfactory, pointers towards an alternative can be found in the literature framed around the reception of Darwin in different countries. Not surprisingly, many of these studies are not actually operating on the scale of the nation at all, but are far more specifically located in terms of group and place. The key subject turns out not to be the overall response in Uruguay; it is the newspaper reports of breeders who discuss the future of their work in relation to Darwinian notions of selection and variation. It is not the debate throughout Mexico; it is eugenic societies in Mexico City and their interventions in public controversies about racial progress. It is not generalised idealist and materialist conceptions of Nature; it is the way these issues are debated at a Freethinkers' Hall in London.

This produces a different problem. As these examples suggest, there is no 'national' reception, but rather an intricately connected series of localised encounters and exchanges. Yet once we abandon the nation-by-nation approach, and recognise the futility of identifying a universally applicable conceptual scheme to capture the essence of 'Darwinism', what will hold these fascinating diverse local studies together?

One possible way forward is to approach evolutionary texts in their concrete material forms: as books, journals, letters, songs, scientific articles, cartoons, illustrated newspapers, public lectures and reports of conversation. We need to think about the availability of these works and their audiences: about reviews, translations, textbooks and elementary redactions. This also means we need to understand Darwin, as well as his followers and his opponents, not as disembodied thinkers but as authors engaged in a commercial system of print and acutely aware of its communicative potential. Historians may pride themselves on having escaped the bearded image of Darwin as sage, but it is surprising how deeply embedded the assumptions behind that image remain; indeed, even the sale and marketing of this image was part of a 'business of culture' that had become strikingly transnational.[2] In short, to under-stand the phenomenon of Darwin and Darwinism it makes sense to start with communication.

This is a promising approach partly because the nineteenth century is an age of great transformation in communication. The most famous symbol of the rapidity of communication was the telegraph, which was invented in the 1830s, with the first transatlantic cable laid immediately after the American Civil War. Vital for transmitting information about finance and diplomacy, by the later nineteenth century it became the channel for news stories. Darwin's death in April 1882 was announced in Bombay only 24 hours after *The Times* had published the news in London.[3] Far more significant for scientific and philosophical debate, however, was the postal system, which was revolutionised not only by the introduction of the Penny Post in 1840, but by the sheer rapidity of improved forms of transport. Once steam ships routinely sailed the Atlantic, the crossing was reduced from several weeks to a few days.

Within Britain, the development of the rail network in the 1840s shortened journey times even more dramatically, from days to hours.

Printing was transformed too, from an artisanal craft to a highly mechanised process involving cheap pulp paper, profuse illustration and machine typesetting. The effect was especially dramatic in the periodical and newspaper press, so that by the 1870s the idea (pioneered in America) of a cheap newspaper affordable by lower and middle class readers was becoming commonplace around the world. The result was an unprecedented increase in the availability and affordability of knowledge.

The place of *Origin*

An obvious starting point for tackling the relation between Darwinism and the culture of communication would be the publication of *Origin* in its first edition. But a book has little meaning except in a context of reception, its relationship to existing genres of discussion and the expectations of its audiences. Insofar as *Origin* made an effective intervention, its history is that of its readers' experience. From this perspective, *Origin* continued a discussion of species, natural law, the origins of man, and the nature of matter itself: issues that had been on the agenda in the Atlantic world at least since the early Enlightenment. This long history needs to be interpreted, not as intellectual genealogy, but as the foreground for understanding the situation readers faced in 1859.

Evolution (or transmutation, as it was usually called) had been an integral part of polite philosophical discussion in eighteenth-century Europe, both in the republic of letters and in the salons (Figure 2.2). The treatises of Benoît de Maillet, Baron d'Holbach and their late eighteenth-century inheritors – notably Jean-Baptiste Lamarck – were widely discussed. The works of Erasmus Darwin (Charles's grandfather) were especially popular, with their discursive footnotes on everything from steam engines to the nature of organic life. Talk about transmutation was a persistent theme of urban intellectual discussion, part of the 'High Enlightenment' in Paris, London and other cosmopolitan centres. In these circumstances, philosophy shaded into pornography, and talk about species and generation blended into innuendo and flirtation (Darnton, 1996).

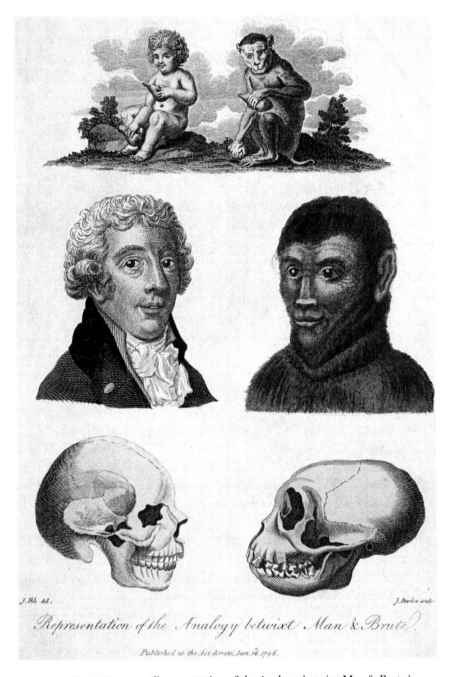

FIGURE 2.2 'Representation of the Analogy betwixt Man & Brute.'
From Sibly, E. (1796). *Magazine of Natural History,* 14 Vols. London: Champante and
Whitrow, **2**, facing p. 151. Author's collection, photograph by R. Horry.

Conversation on these topics became increasingly subdued in the counter-revolutionary paranoia of the early nineteenth century. In Regency and early Victorian England, transmutation was denounced as the property of dissolute Frenchmen, dissenting medical writers and working-class radical secularists. The scientific poems and other writings of Erasmus Darwin, previously among the most fashionable works of the age, were parodied as monstrosities in verse, disfigured by their enthusiastic advocacy of an outmoded philosophy. Evolutionary transformism was effectively placed outside the boundaries of legitimate science: it was speculative, perhaps just silly, but also potentially dangerous and divisive. As Robert Mudie told aspiring working class readers in *Man, in his Physical Structure and Adaptations* in 1838, it was 'the doctrine of materialism in its most malignant and inveterate form' and 'no laughing matter' (Mudie, 1838).

If the science that dominated public life during Darwin's formative years did have a general lesson, it was that of progress. From the 1820s, the emerging science of geology had focused on mapping and determining the order of the strata; but in so doing a spectacular succession of lost worlds had been revealed, showing that low forms of invertebrate life had been succeeded by successively more advanced fish, reptiles, birds, mammals and humans. Similarly, contemporary medical works, particularly those imported from the continent, used the development of the embryo to illustrate a series of advances towards a human form. The same progressive message was emblazoned on the heavens. At Birr Castle in Ireland, Lord Rosse's gigantic reflecting telescopes – the largest ever built up to that time – pointed to evidence for cosmic advance on the grandest scale (Figure 2.3). As the astronomer John Pringle Nichol wrote in 1837, 'In the vast Heavens, as well as among phenomena around us, all things are in a state of change and PROGRESS; there too – on the sky – in splendid hieroglyphics the truth is inscribed, that the grandest forms of present Being are only GERMS swelling and bursting with a life to come!' (Nichol, 1837).

However evocative they might be, these spectacles of geology, astronomy and physiology were not evolutionary. They were progressive and sequential, but the scenes they depicted remained separate and distinct, like individual images in a contemporary phantasmagoria or rolling

FIGURE 2.3 The great spiral nebula. From Nichol, J. P. (1850). *The Architecture of the Heavens.* London: John W. Parker, plate 12, facing p. 75. Author's collection, photograph by R. Horry.

panorama. Transmutation, especially in relation to life, remained on the fringes of public debate in the English-speaking world; it was the sort of subject reserved for select male conversation once the women had left the room.

This situation changed dramatically with the publication in October 1844 of *Vestiges of the Natural History of Creation*, a hugely ambitious work that attempted to bring all the phenomena of Nature under an overarching law of development. As readable as a novel, *Vestiges* began with nebular condensation, continued through the formation of the solar system and the rise of life on Earth, and concluded with racial progress and the moral and spiritual destiny of man (Figure 2.4). Written anonymously, the book was widely translated, reprinted and read across the social and religious spectrum in Britain and the colonies, the rest of Protestant Europe, and the United States. Although the author argued

FIGURE 2.4 Jokes about transmutation – in this case, the racial
development of Chinese labourers as a result of success in the Australian gold rush of
the 1850s – were common well before Darwin's *Origin of Species*. From 'New illustration
of progressive development'. Respectfully dedicated to the author of *Vestiges of
Creation*. *Melbourne Punch*, 3 July 1856.

that God worked through law rather than miracle, he or she was often
accused of outright atheism. Even so, *Vestiges* brought the question of
evolution on to the public agenda for the first time since the eighteenth
century. It would become the one book with which all readers of *Origin*
could be expected to be familiar (Secord, 2000).

Although a few practitioners supported the idea of linking the sciences
in a causal evolutionary account, among genteel circles such ideas were
usually thought to be too ridiculous to deserve discussion. This view is
reflected when, only months before *Vestiges* was published, Darwin told
Joseph Hooker about his secret work on species origins. The letter is
pervaded by an acute sense of embarrassment; as Darwin half-jokingly
remarks, it was 'like confessing a murder'. He had no fear of murdering
God, Christianity, Victorian society: Darwin was worried about his
scientific reputation. 'You will now groan, & think to yourself, "on what
a man have I been wasting my time in writing to." '[4] Worthless speculation: this was pretty much the way that evolution continued to be viewed
by respected men of science in the 1840s and 1850s.

On its publication in 1859, what *Origin* did was to make the study of
evolution safe for scientific research. Ironically, the really novel claim

in the book was largely rejected, so that the mechanism of natural selection became central only with the rise of the modern evolutionary synthesis in the 1930s, half a century after Darwin's death. Instead, *Origin* made it possible for men of science to talk about evolution as part of their regular work without having to be embarrassed or equivocal. They could stop having to confess murders. We can see the importance of this in what Darwin himself did immediately after *Origin*; for rather than publishing another big speculative book, or the first instalment of his promised expansion and elaboration of his evolutionary theory, he wrote on the fertilisation of orchids. In effect, he told his fellow naturalists that they could do serious science with this new evolutionary approach. It could serve as the basis for new experiments, observations, publications and careers. Looking back at the 20-year gap between *Origin* and the initial formulation of natural selection in 1838, we can see why Darwin had taken so long to construct his theory. He wanted all the supporting evidence and argument to be in place, so that contemporaries could be convinced that a scientific way of going forward could be found in a theory of evolution. Darwin wanted his revolution to be a fait accompli, total and irrefutable.

As much as Darwin cared for the fate of his explanatory mechanism of selection, the main point was getting a natural view of human, plant and animal origins onto the agenda of science. This proved crucial in terms of the reception of Darwin's work by other men of science. The naturalist and anatomist Thomas Henry Huxley, who had opposed transmutation specifically because of its 'unscientific' character, took *Origin* immediately to his heart. Surprisingly, Huxley did not believe that natural selection was the correct mechanism, but he did argue that it was (to use a phrase later coined by Darwin) a 'theory by which to work'.[5] Above all, Huxley saw in Darwin's book a way of giving practising men of science their rightful place on an intellectual stage too long arbitrated by theologians, philosophers and classicists. From this perspective, this was the birth of a new reformation, in which real knowledge would rip through the cobwebs of tradition. As Huxley wrote of *Origin* in the *Westminster Review*, 'Extinguished theologians lie about the cradle of every science as the strangled snakes besides that of Hercules' (Huxley, 1860).

A scientific progress

'Darwinism' served in these debates as a flag to signal the ambitions of a new generation of men of science in Europe and North America. In introducing the term in the *Westminster*, Huxley naturally meant to associate it with his own reading of Darwin's views.[6] Like most 'isms', however, 'Darwinism' always resisted definition, and with good reason. It offered no core set of ideas, doctrines or system of thought, but instead an arena for discussion, often involving issues never mentioned in Darwin's writings. It is best, therefore, not to start with Darwinism as a doctrine, but rather with the debates, dialogues, conversations and controversies surrounding the circulation of particular works. In this sense, *Origin* and *Descent* are fault-lines along which much larger cultural changes were expressed and worked out.

The Darwinian debate employed all the means of communication, new and old, that were available in the later decades of the nineteenth century. For example, on the Indian subcontinent, where Darwinian evolution featured in discussions of Bengali independence, most discussion occurred orally. Similarly, the crucial forums in London's West End were conversational, at dinner parties and scientific soirées. Beyond this, the key arenas for discussion were weekly newspapers and literary periodicals, particularly those which encouraged their readers to engage with the latest work in science. The subjects Darwin had opened up were precisely the sort that these new publications were designed to cater for. Such periodicals only rarely developed inflexible party positions about science; instead they served to make Darwinism a field for interaction, debate and discussion.

Publications of this kind were becoming characteristic features of modern urban life not just in Europe and the United States but throughout the world. In this context of contemporary journalism, evolutionary discussion called upon scientific understandings of the past, as unearthed by geology and astronomy, but the questions it asked were about the future. What implications did the sciences hold for tradition, in a world of industrial development and imperial expansion? Darwinism became a central arena for considering these wider issues and concerns. At stake was the fundamental question of what it meant to be modern.

FIGURE 2.5 'The coming man' and 'The coming woman', *The World*, New York, 12 March 1871. Copy at CUL, Add. mss DAR 140.1.1. Reproduced by permission of the Syndics of Cambridge University Library.

Take, for example, the bizarre images that greeted readers on the front page of the New York *World* in March 1871.[7] Part of an essay mocking Darwin's *Descent of Man*, these showed 'The coming man' and 'The coming woman' (Figure 2.5) as unnatural outcomes of natural selection. After all, as one of the participants in the accompanying dialogue noted, what could be expected when the institutions of society protected the weak and feeble-minded? These super-evolved humans, living in an era of all play and no work, would have very big heads and very small muscles. Chemically produced artificial foods would lead to a hugely modified digestive system and the slimmest of waists. Men would lose their teeth, women their breasts; and everyone would be able to smoke opium and drink alcohol without adverse effect.

Another possible future had been imagined a few years earlier in *The Press*, a daily newspaper issued in Christchurch, New Zealand. In 1863, an anonymous essay 'Darwin among the machines' asked what would happen now that machines were evolving more rapidly than humans. 'Day by day', the author claimed:

> the machines are gaining ground upon us; day by day we are becoming
> more subservient to them . . . Our opinion is that war to the death
> should be instantly proclaimed against them. Every machine of every
> sort should be destroyed by the well-wisher of his species. If it be urged
> that this is impossible under the present condition of human affairs,
> this at once proves that the mischief is already done, that our servitude
> has commenced in good earnest, that we have raised a race of beings
> whom it is beyond our power to destroy, and that we are not only
> enslaved but are absolutely acquiescent in our bondage.[8]

This nightmarish scenario, prompted by the publication of *Origin* four years earlier, was the work of an émigré sheep farmer, the young Samuel Butler, who would soon return to Britain to make it the basis of his utopian novel *Erewhon* (1872). In the next few decades, H. G. Wells, Camille Flammarion, Jagadananda Roy and others made the new genre of science fiction into a lasting monument of the Darwinian debate.

These examples, drawn from the opposite ends of the world, illustrate the way in which Darwin's writings afforded fresh opportunities to contemplate the future. Where are we going? What might happen to us? What will it mean to be human? Reading and talking about these issues was what 'Darwinism' became: a shared context to work through highly local, specific experiences in relation to transformations across the world. For readers of the Christchurch or New York press, those changes would have been identified primarily with colonial expansion, trade and the progress of empire. They are exemplified by 'American Progress', painted by John Gast in 1872, just a year after publication of *The Descent of Man*, and widely distributed as a popular chromolithograph (Figure 2.6). A diaphanously clad woman with the star of empire on her forehead floats above a varied landscape. In her right hand is a school book, and accompanying her are all the wonders of modern technology, from the stage coach to the wires of the telegraph she lays down with her left hand. Their light puts into shadow the receding world of the buffalo, bear, wild horse and native American. This association between new communications technologies and racial progress was central to defining Darwinism.

FIGURE 2.6 'American Progress', chromolithograph (c. 1873) by George A. Crofutt after oil painting by John Gast, 1872. Library of Congress.

This is evident in western attitudes to the Boxer Uprising in China between 1898 and 1901. Supported by the Qing Dynasty, which had lasted for a quarter of a millennium, the Boxers were accused of opposing all change, leading European and North American commentators to claim that the Chinese were ignorant, intolerant and against progress. As a consequence of the advocacy of evolutionary progress through a struggle for existence, as defined by many in the Darwinian debates, the Chinese were depicted with racially inferior low foreheads and vacant expressions. (Not surprisingly, in stressing the virtues of civilisation, little attention was paid to the brutal methods used by the Allied Powers in forcing modernisation upon the country.)

Within China, reformers attempted to turn this application of social evolution back upon the enemy, using the pre-eminently modern technology of the newspaper. The leading proponent of Darwinian debate in China was Liang Qichao, who not incidentally also established seven different newspaper titles, including *Journal of a New People* in 1902. To escape censorship, the paper was published in Japan and smuggled into the treaty ports. The polemical, partisan journals edited by Liang became leading forums for evolutionary discussions, aimed at introducing the views of Darwin, Huxley and their followers so that the Chinese people could understand the need for reform. The old regime, it was argued, with its support for traditional ways of life, had brought the country to the point where the Chinese were evolving backwards. As Liang wrote:

> Our country's ignorant masses, four hundred million of them, have for thousands of years been kept under control by a people-ravaging government, so that now they are like blind fish born in a black cavern, who come out into the ocean and cannot see . . . But people with a slave mentality are not just content to be slaves themselves, they insist on ridiculing those who are not slaves. Alas, to pit such people against the races of Europe in this world of struggle for survival and survival of the fittest – What hope is there? What hope is there?[9]

Reformers were encouraged to confront the forces of the west as part of the racial struggles of Social Darwinism. Naked aggression, once thought barbaric, was now presented as a law of civilisation supported by European and American science. In Liang's view, the power of the press was almost unlimited: 'The newspaper gathers virtually all the thoughts and expressions of the nation and systematically introduces them to the citizenry, it being irrelevant whether they are important or not, concise or not, radical or not. The press, therefore, can contain, reject, produce, as well as destroy, everything'. Only by an evolutionary understanding conveyed through the mechanisms of modern journalism, it was argued, could the Chinese people understand their proper role in the world. 'How great is the force of the newspaper!' Liang claimed, 'And how grave is the duty of the newspaper!'[10]

Evolutionary controversy and international communication were equally entangled in the Arabic-speaking world. The leading periodical for these discussions, established in Beirut in 1882, was *al-Muqtataf*, which became a lively forum for debating the consequences of evolutionary views (Figure 2.7) (Glaβ, 1994; Elshakry, 2007). An article on 'Darwinism' expressed a moderate view, in contrast to the evangelical

FIGURE 2.7 Title page of *al-Muqtataf*, Beirut: 1882. Library of Congress.

Christian missionaries who had unceremoniously ejected the Harvard-educated geology professor from the Protestant College in Beirut who had dared to eulogise Darwin. As the author noted:

> Whatever mistakes and missing links there are in Darwin's theory or whatever errors were added to it, there is no doubt that despite these limitations, it now includes established truths and that it has given scientists many benefits and opened for them paths to [uncover] unsolved problems in a number of ways. And so it should be said that the just will be pleased with the truth wherever they see it and accept it as a gift from the Lord however it comes.

Similarly, the writings of the Lebanese legal scholar and author Husayn al-Jisr, commenting on *sharī'ah* law, had acknowledged that if the truth of evolution could be proved absolutely then passages in the Koran would need to be interpreted accordingly.[11]

Tellingly, it was a leading advocate of economic modernisation who was most concerned about the degrading implications of evolutionary philosophy. In *Refutation of the Materialists* (1881) the Persian author Jamāl al-Dīn al-Afghānī argued that the Islamic world urgently needed technological and industrial innovation from the west, but that materialism, and Darwin's views in particular, had to be rejected. His initial response to *Origin* was entertainingly simple, refuting the book by complaining that a flea could never evolve into an elephant. But al-Afghānī eventually softened his position, acknowledging that Darwin had rejected the more extreme claims of the materialists.[12]

What is striking, even from these few examples, is the range and sophistication of discussion and debate. Darwinism did not simply reinforce racial hierarchies and the aims of empire, nor was it simply introduced from the west as a static ideology. Instead, from Beirut to Beijing, Darwinism offered ways of changing the boundaries of discussions and making sense of local disagreements. The debate became, for those on all sides, part of worldwide transformation of communication in the late nineteenth and early twentieth centuries.

The range of that discussion is apparent from the chart (Figure 2.8), which shows when *Origin* was first translated into different languages. German and French were crucial, as these could be read by learned elites throughout Europe and the rest of the world. The initial wave of activity

- German 1860
- French 1862
- Dutch 1864
- Russian 1864
- Italian 1864
- Swedish 1869
- Danish 1872
- Polish 1873
- Hungarian 1873–74
- Spanish 1877
- Serbian 1878

- Japanese 1896
- Chinese 1903
- Czech 1914
- Latvian 1914–15
- Greek 1915
- Arabic 1918

- Portuguese 1920s
- Finnish 1928
- Armenian 1936
- Ukrainian 1936
- Bulgarian 1946
- Romanian 1950
- Slovene 1951
- Korean 1957
- Flemish 1958
- Lithuanian 1959
- Hebrew 1960
- Hindi 1964
- Turkish 1970

FIGURE 2.8 Translations of Darwin's *Origin of Species*, compiled from the Freeman Bibliographical Database at http://darwin-online.org.uk, revised by J. van Wyhe from Freeman, R. B. (1977). *The Works of Charles Darwin: An Annotated Bibliographical Handlist*, 2nd edn. Folkstone, Kent: Dawson, pp. 73–111.

ended in the late 1870s. There was then a gap before translations started to become available in other parts of the world, notably in Japanese, Chinese and Arabic. In this phase, the debate became much more international. The third column shows a pattern of translations into the minor European languages during the Soviet era, when Darwinism was supported as part of the intellectual legacy of Marxism.

Translating Darwin's writings was never straightforward. Many of *Origin*'s central terms ('natural', 'selection', 'race', 'descent', 'struggle', 'fittest') were immediately subject to wildly varying meanings in English, let alone in different linguistic settings. Darwin and his friends condemned some translations as irredeemably bad, including Clémence Royer's crusading materialist version in French; but in fact the problem of translations was simply part of an intractable problem of communicating the essence of a complex argument in one language into another. And of course, translators had their own agendas. In Arabic, as Marwa Elshakry has shown, 'Darwinism' was usually translated as '*madhhab Dārwīn*', giving the new views a particular authority, as '*madhhab*' was associated with an orthodox school of Muslim law, and also with philosophical system (Elshakry, 2008). In German, 'Darwinismus' had militantly anti-religious and political connotations, and links with Ernst Haeckel's views on embryonic recapitulation (Daum, 2002).

There can be little doubt that the vast majority of readers, despite the remarkable range of other translations, approached Darwin's text either in English, German or French. The importance of further translations was often symbolic, usually (as in the case of Russian) as a gesture of defiance against religious or political authorities. Most of those who wished to read a scientific book of some five hundred pages were likely to be in command of one of the leading learned languages already. Thus *Origin* continued to be read in Spain and Latin America primarily in French long after a Spanish translation appeared in 1877 (Glick, 1974). In any event, even among the educated, the discussion centred on newspapers and periodicals rather than full-length books, and the exposure of readers even to these was far from universal. Only a small part of the world's population, mostly in urban and intellectual centres, had heard of Darwin or Darwinism; and the number of those who read Darwin's works, even in translation, was inevitably a tiny proportion of that minority.

Darwinism was never a pervasive or totalising ideology, but the outcome of criss-crossing lines of communication between diverse and often scattered localities. Only in that sense has Darwin ever been 'global'.

Darwinian questions

The bicentennial outpouring of interest in Darwin is a continuation of precisely the sort of debates I've been talking about in Europe, the Middle East, the United States, China and other parts of the world. There have of course been dramatic changes, not only in wider global politics, but in the politics of scientific disciplines. Notably, natural selection has gained a new order of importance outside its original strongholds in ecology, biogeography and the biology of whole organisms. In the nineteenth century, most scientists would have said that natural selection was a dead end, while acknowledging the general contribution of Darwin to philosophy and our self-understanding. Even after the modern synthesis of the 1930s made evolution relevant to laboratory biology, its connection to some of the most prestigious areas of the life sciences, such as biochemistry and molecular biology, remained

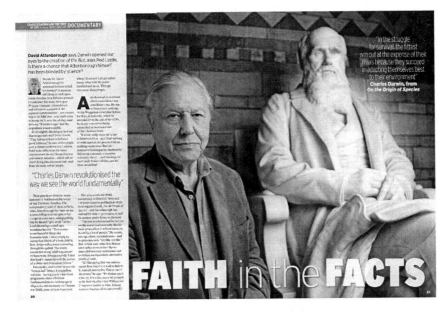

FIGURE 2.9 R. Liddle, Faith in the facts. *Radio Times*, 31 January 2009, pp. 20–1. Reproduced by permission.

tenuous at best. During the past 20 years, largely through the rise of genomics and embryological genetics, this situation has changed dramatically. At the same time, the rise of evangelical fundamentalism, from Istanbul to Indiana, has given the defence of evolution a rightful urgency that it has never had before. These factors combine to make the celebration in 2009 unparalleled in extent and significance.

Darwin thus has more symbolic value than ever; yet we have lost much of the sense of dialogue and careful discussion that made the earlier debates so rich. Take this spread (Figure 2.9) from the *Radio Times* in January 2009, showing the broadcaster and naturalist David Attenborough at the Natural History Museum in London. Despite the interest of the article, which included a fascinating and controversial interview, there is a problem with the quotation in the upper right hand corner, purportedly from *Origin*:

> In the struggle for survival, the fittest win out because they succeed in adapting themselves best to their environment.

This is something Darwin never said and (as any elementary biology student ought to know) never would have said, for it gives too much stress to self-willed environmental adaptation. In fact Darwin looks rather alarmed by it.

Why has this mistake been made? This is neither the fault of Sir David Attenborough nor of the *Radio Times* (media people often have an unjustifiably bad name in these matters), but of the Natural History Museum in London, where these pictures were taken, and whose website prominently features this quotation, even identifying it as from *Origin*. With all of Darwin's writings freely available on the internet, it is easy to check in a few seconds.[13] The quotation crops up most frequently in management websites and gives a biological basis to the encouragement of teamwork in that context. It would appear that one of the world's greatest natural history museums, which holds many of Darwin's specimens and employs many experts on Darwinian matters, is no longer a dependable source of information. The experts are there, but all too often they have little say in what is put before the public.

An even more telling, and certainly more expensive, example involves the California Academy of Sciences. In their recently opened headquarters in San Francisco, which cost half a billion dollars, the stone in the floor has embedded within it the following quotation, attributed to Darwin:

> It is not the strongest of the species that survives, nor the most intelligent that survives, it is the one that is most adaptable to change.

Now it could be argued that the previous quotation had Darwinian aspects, but this one simply forges a new Darwin, one that stresses adaptability over fitness in a way that is antagonistic to most current understandings of his writings. There is a self-reflexive irony here, for this widely distributed quotation unintentionally celebrates our extraordinary ability to adapt Darwin to new circumstances. As a translation into the idiom of contemporary corporate management, it is the latest chapter in a process of writing and rewriting that has gone on for the past century and a half.

You can blame the California Academy of Sciences, the Natural History Museum, management consultants or the internet, but they simply point to a deeper source of the problem. Darwin's writings don't give easy sound bites. My own experience in searching for an appropriate quotation for a publisher to attract potential readers to an anthology bears this out. There are half a dozen bits of prose in the whole Darwinian corpus that are simultaneously poetic and punchy; and in the end, it proved impossible to find a decent quotation that hadn't been used to death. Now that says something interesting: that Darwin is not very quotable, and in fact is not easy to engage with unless through careful contemplation of long stretches of argument. Darwin's texts are complicated, rich and ambiguous. They do not give simple doctrines.

Take the basic principles of evolution as laid out in *Origin*. Although it might seem possible to sum these up in a few words, Darwin's own explanations are neither simple nor straightforward, which has led to long-running disputes about the level of organisation to which natural selection is supposed to apply. Is it the species? The group? The individual? The particular character? Darwin's answers might tend towards certain positions, but there is some evidence for all of these views in his writing. At an even more basic level, *Origin* nowhere claims that natural selection is the only mechanism for evolutionary change, but gives room to the kind of inheritance through use often associated with Lamarck.[14] Yet eminent biologists and philosophers often act as though Darwin's model can be reduced to a handful of logical propositions, and that any deviation from these was introduced in late editions as an ill-judged concession to ignorant critics.

Even more famously complex and indeterminate are the theological foundations of Darwin's writings. Contrary to widespread assumptions, these appear to assume the existence of God; for example, *Origin* refers to 'the laws impressed on matter by the Creator'. Lest readers fail to get the message, all editions after the first add an even more direct reference in the final sentence, which speaks of 'life being originally breathed by the Creator into a few forms or into one'. That is what Darwin wrote. Such phrases remained in the text to the end of his life, long after he claimed in a letter to Hooker to have 'long regretted that I truckled to public opinion & used Pentateuchal term of creation, by which I really meant "appeared" by some wholly unknown process. −'[15] This letter has often been quoted by freethinkers and atheists (including Daniel Dennett and Richard Dawkins), who use it to reduce the complexity of *Origin* to a crude materialism. In fact, references to the Creator are neither sops to believers, nor straightforward claims for miraculous intervention.

These intractable dilemmas in getting at what Darwin 'really meant' point up something about the way he writes. His sentences are complicated. His books are long. They open up questions and opportunities for debate, discussion and conversation. We remember Darwin not just because he had a great idea, but because of the multiple ways that his writings can be read, because of their generosity, openness and experience. These are reasons his works are debated in laboratories, museums, courtrooms, websites, school classrooms and television documentaries. They are important because of – not in spite of – seeming contradictions and ambiguities.

The person who saw this most clearly was the American philosopher John Dewey, speaking almost exactly 100 years ago at a Darwin centennial gathering at Columbia University in New York. What Darwin offers, Dewey recognised, are not answers to the dilemmas posed by science, but new questions. That, ultimately, is the key to the continuing relevance of his works in so many different situations across the world.

> Old ideas give way slowly; for they are more than abstract logical forms and categories. They are habits, predispositions, deeply engrained attitudes of aversion and preference. Moreover, the conviction persists – though history shows it to be a hallucination – that all the questions that the human mind has asked are questions that can be answered in terms of the alternatives that the questions themselves present. But in fact intellectual progress usually occurs through sheer abandonment of such questions, together with both of the alternatives they assume – an abandonment that results from decreasing vitality and interest in their point of view. We do not solve them: we get over them. Old questions are solved by disappearing, evaporating, while new questions corresponding to the changed attitude of endeavor and preference take their place. Doubtless the greatest dissolvent of old questions, the greatest precipitant of new methods, new intentions, new problems, is the one effected by the scientific revolution completed in the 'Origin of Species'.
>
> (Dewey, 1909)

Acknowledgements

I am grateful to my colleagues at the Darwin Correspondence Project, particularly Paul White, for suggestions and comments. Catharine Hall provided the opportunity to present an early version at her seminar 'Reconfiguring the British: Empire, Nation, World'. I am particularly grateful to Anne Secord for reading drafts of this essay.

3 Darwin in the literary world

REBECCA STOTT

For twenty years or so, for various reasons, my attempts to understand metaphysical questions have been tangled up with trying to understand Darwin and what his ideas mean in the widest sense. My writing in fiction and non-fiction has been preoccupied with Darwinian ways of seeing. I published a book on Darwin's barnacle years in 2003. My first novel *Ghostwalk* turns on the idea of entanglement conceived of in implicitly Darwinian ways. My most recent novel, *The Coral Thief*, set in Paris in 1815, is a coming-of-age story about a young medical student who, falling in amongst a gang of infidel thieves, comes to see the world through Lamarckian eyes. For many contemporary writers, Darwin's ideas provide endless forms for contemplation, translation and transformation.

Why might it be interesting to investigate Darwin's impact on the literary world? There are several reasons. The appearance of Darwinian ideas in European novels and poetry from the 1860s onwards shows us the extraordinary diversity of philosophical and literary responses to Darwin's ideas. That diversity of response reminds us that texts, even scientific texts like *On the Origin of Species by Natural Selection*, are like kaleidoscopes: different people see different things at different times. And, if we didn't already know this, it reminds us that novelists and poets are not just disseminators – they are translators and interpreters; they don't just replicate ideas, they *transform* them; they show us how complex ideas can be adapted for new ways of seeing or understanding.

But before we set off down that path, it might be useful to know what kind of books Darwin liked, for he was a reader as well as a writer. In his

Darwin, eds. William Brown and Andrew C. Fabian. Published by Cambridge University Press. © Darwin College 2010.

autobiography he tells us what he liked to read. He read, he tells us, Horace's *Odes* at school; this was almost the only poetry he could memorise and remember for more than a day. He read the Romantic poets with pleasure: Wordsworth and Coleridge and the works of Byron. He read Shakespeare in the window seat of his school and Jane Austen on *HMS Beagle*; he and FitzRoy had a shared love of Austen. It is tempting to think of the two men arguing about Emma or Mr Knightley whilst the ship, loaded with bottled and stuffed specimens, sailed through darkened seas. We also know that he re-read Milton's *Paradise Lost* on the *Beagle*. Gillian Beer has shown that it helped shape his sense not only of the fecundity of nature but also the complex interdependence of good and evil.

Between the late 1830s and the 1850s Darwin kept reading notebooks scoring the books he read or which Emma Darwin read to him as 'good', 'curious', 'average', or 'very good' and, occasionally, 'excellent'. From these notebooks we know that he read widely and enthusiastically and that he liked Thomas Carlyle and Harriet Martineau, Charles Dickens and Elizabeth Gaskell, travel narratives and biographies. Whilst he claims in his autobiography that he lost his taste for poetry, music and painting in his latter years, he explains with relief that he did not lose his pleasure in novels:

> On the other hand, novels which are works of the imagination, though not of a very high order, have been for years a wonderful relief and pleasure to me, and I often bless all novelists. A surprising number have been read aloud to me, and I like all if moderately good, and if they do not end unhappily – against which a law ought to be passed. A novel, according to my taste, does not come into the first class unless it contains some person whom one can thoroughly love, and if it be a pretty woman all the better.
>
> (de Beer, 1974, p. 84)

So I sometimes wonder what Darwin might have made of the fact that so many novels published since 1860 are woven through with Darwinian ideas. Given how much he liked them, he might have been saddened by the fact that very few profoundly Darwinian novels have happy endings. The logic of the Darwinian story does not allow for happy endings; it doesn't really allow for an ending at all. What would Darwin have thought about Hardy's *Tess of the D'Urbervilles*, for instance, published in volume form in 1891, nine years after Darwin's death? Tess is a pretty girl. She is a very pretty girl. Darwin would have liked that. But that is why she can't have a happy ending, Hardy insists. She is hounded by the forces of natural and sexual selection, punished for her own sexual desires, punished because she is desired, punished because chance seems to turn against her at every corner and she struggles to survive, moving relentlessly towards that final sacrifice: her last night spent sleeping on the stones of Stonehenge before she is taken away to be tried and hanged. Her death is a kind of ritual slaughter, orchestrated by an author whose encounter with Darwinism confirmed his own conviction of both the beauty and the mindless, chaotic destruction of the world.

Would Darwin have passed a law against such a book, against such a way of seeing? He might have. And there's an irony in that. Others have come to describe Hardy's novels and poetry as expressing 'the ache of modernism', deploying a phrase Hardy coined in *Tess of the D'Urbervilles.* If it is an ache it is a *Darwinian* ache.

Here is a selected list of writers who have either been directly or indirectly influenced by Darwinian ideas:

Charles Kingsley	Franz Kafka
Robert Browning	T. S. Eliot
Gerard Manley Hopkins	Edward Thomas
W. B. Yeats	Elizabeth Bishop
Robert Louis Stevenson	John Updike
D. H. Lawrence	George Eliot
James Joyce	Samuel Butler
Dylan Thomas	Thomas Hardy
Amy Clampitt	H. G. Wells
Ian McEwan	Robert Frost
Lewis Carroll	Virginia Woolf
Matthew Arnold	Ted Hughes
George Meredith	Wallace Stevens
H. P. Lovecraft	A. S. Byatt

Lists are suggestive but it's the detail and the texture of these encounters between writers and Darwinian ideas that is interesting. We need to step closer. We need to look at individuals and particulars, some epiphanies and some moments of mutual recognition. And we need to go back to the beginning, to a rectory in Hampshire.

Endless forms

In August 1859, Darwin sent out 80 presentation copies of *On the Origin of Species by Natural Selection* to the men in Britain and Europe who, in his view, would shape its reception. That list included men of science, geologists, churchmen, historians and medical men, but it only included one novelist. In November 1859, the postman delivered a carefully wrapped presentation copy of Darwin's book to a quiet rectory in Eversley in Hampshire. Four days before the book went on sale, the rector, who was a keen biologist, wrote the following letter to its author:

> Eversley Rectory, Winchfield,
> November 18th, 1859.
>
> DEAR SIR, – I have to thank you for the unexpected honour of your book. That the Naturalist whom, of all naturalists living, I most wish to know and to learn from, should have sent a scientist like me his book, encourages me at least to observe more carefully, and think more slowly.
> I am so poorly (in brain), that I fear I cannot read your book just now as I ought. All I have seen of it awes me; both with the heap of facts and the prestige of your name, and also with the clear intuition, that if you be right, I must give up much that I have believed and written.
>
> (*The Correspondence of Charles Darwin*, VII, p. 379)

And then the rector went on to tell Darwin two things – that, as a biologist, he had long since 'learnt to disbelieve the dogma of the permanence of species'. Secondly, he told Darwin that he was sure that God made all the first forms with an ability to evolve into all forms needful. Then he signed himself Charles Kingsley.

Three years later Kingsley began to write a book published in 1864 as *The Water Babies*. In it, Tom, a child chimney sweep on the run from his master, falls into a river and, with the help of some fairies, turns into a water baby. Here is a taste of Kingsley's delightfully playful and Rabelaisian prose:

> No water-babies, indeed? Why, wise men of old said that everything on earth had its double in the water; and you may see that that is, if not quite true, still quite as true as most other theories which you are likely to hear for many a day. There are land-babies – then why not

water-babies? Are there not water-rats, water-flies, water-crickets, water-crabs, water-tortoises, water-scorpions, water-tigers and water-hogs, water-cats and water-dogs, sea-lions and sea-bears, sea-horses and sea-elephants, sea-mice and sea-urchins, sea-razors and sea-pens, sea-combs and sea-fans; and of plants, are there not water-grass, and water-crowfoot, water-milfoil, and so on, without end?

(Kingsley, 1864, p. 70)

Tom undergoes a series of extraordinary metamorphoses, learning import-ant moral lessons along the way, until he emerges finally as an English gentleman, returns to land and marries the pretty girl of his dreams. Yes, the novel has a happy ending. The story is a coming-of-age narrative like so many Victorian novels (think of *Oliver Twist, Great Expectations* and *Jane Eyre,* for instance), but this is the only coming-of-age story I know of that is set underwater. Halfway through his journey Tom comes across the sea mother at the end of the world who is making endless forms which she sends out into the world to make themselves. So within five years of the publication of Darwin's *Origin,* Charles Kingsley had turned Darwinian evolution into a moral fable of self-making and redemption (Figure 3.1).

There are certain striking features of *The Water Babies* that are shared by other first generation literary interpreters of Darwin's work. Kings-ley's novel *celebrates* human–animal kinship and biological mutability; it is not disturbed by it. He shows us an undersea world in which endless forms have been and are being created. Secondly, Kingsley shrinks Tom in order to make him smaller than the other animals he encounters. He must shrink, it seems, in order to grow. Thirdly, Kingsley makes the aquatic the source and site of the biological mutability of the story. This is an important point because literary critics have tended to concentrate on the primate as the evolutionary icon, the key and iconic expression of a fear of 'a beast within'. However, if you look at the full range of nineteenth-century literary texts that engage with Darwinian ideas, and particularly the first generation novels and poems, the feared Dar-winian ancestor or monstrous primary form or 'beast within' is just as often imagined as a sea creature of some kind, something slippery and amphibian. These Darwin-inspired sea monsters have been overlooked in the search for the ape monsters. Think of Robert Browning's 'Caliban on Setobos', for instance, published in 1864, or consider Jane Carlyle's

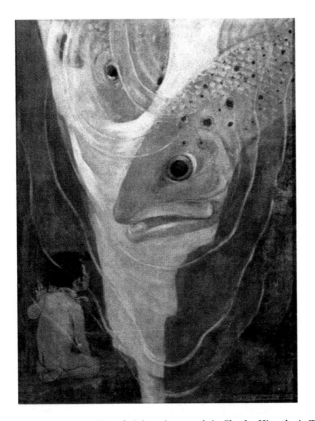

FIGURE 3.1 Tom shrinks to be remade in Charles Kingsley's *The Water Babies.*

delightfully sarcastic account of reading *On the Origin of Species,* expressed in a letter to a friend in 1860 in which she lampoons evolution:

> But even when Darwin in a book that all the scientific world is in ecstasy over, proved the other day that we are all come from shell-fish, it didn't move me to the slightest curiosity whether we are or not. I did not feel that the slightest light would be thrown on my practical life for me, by having it ever so logically made out that my first ancestor, millions of millions of ages back, had been, or even had not been, an oyster. It remained a plain fact that I was no oyster, nor had any grandfather an oyster within my knowledge; and for the rest, there was nothing to be gained, for this world, or the next, by going into the oyster-question, till all more pressing questions were exhausted!
>
> (Jane Carlyle, letter to Mrs Russell, 28 January 1860)

Kingsley, in his novel, like other writers who followed him, reflects what we might call the chiaroscuro of Darwinism – the coexistence or shadowing of joy and cruelty, reproduction and destruction. Everywhere behind the talking animals and mutating human forms and the comedy and the playing around with scale, Kingsley reminds us of the cruelties in the struggles for existence. He won't let us forget that. He won't let us look away. But he will give us – and Tom – a happy ending.

What would Darwin have made of *The Water Babies?* Though we know he was delighted by Kingsley's approving letter, I think he might have found the novel a little embarrassing. All that material about the sea mother and the fairies would have been too mystical for Darwin's taste, I suspect, and the trajectory of the book far too teleological. Souls make bodies, Kingsley insists at the heart of this book. Darwin, who had long since drifted away from belief in souls, would have called that nonsense (though not in public, of course).

There were other writers in the 1860s preoccupied with the implications of Darwin's ideas. By 1862, Charles Dodgson, Oxford don and mathematician, had no doubt grown pretty tired of high-table conversations about Darwin. He much preferred the company of children, and particularly the company of the daughters of the Dean of Christ Church, the Liddell girls. He took them out rowing a good deal in the summer of 1862 and on one particular day he told the girls a story about a girl falling down a rabbit hole. Two years later he wrote it all out by hand with illustrations and gave it to Alice Liddell as a present and entitled it *Alice's Adventures Under Ground.* A year later, in 1865, he published an extended version illustrated by John Tenniel and entitled *Alice's Adventures in Wonderland.*

Alice's Adventures in Wonderland is the thought experiment of a philosopher fascinated by questions of logic in particular, but it is also shot through with Darwinian ideas. Alice, like Tom, *falls* into an alternative world which operates according to bizarre and chaotic rules but which is also full of mutating and expanding and contracting animal and human body parts. A girl's body stretches, contracts and elongates in an underground world of eat or be eaten in which an endless variety of hybrid and grotesque animals of all sizes prey on each other, including a Cheshire Cat with a grin like a crocodile and a human baby that turns into a pig, a doomed Bread and Butter fly, a Mock Turtle, a Gryphon and the most famous extinct animal of all, the Dodo. At one point of the narrative, Alice falls into a pool of her own tears and swims to safety alongside a host of other winged and feathered and furred creatures. Like Kingsley, Carroll turned fragments of Darwinism into Rabelaisian *endless forms*, animal–human hybrids that shrink and mutate and hybridise. A hookah-smoking caterpillar sitting on a mushroom stool asks Alice: 'Who are you?' That was the key Darwinian question of the 1860s. Who are you? What is a man? Where does a human begin and end (Figure 3.2)?

This generation of mutating animal–human hybrids produced in the wake of Darwin's book may have been charming and benign in the hands of Carroll and Kingsley, but there were malign appearances of Darwin-inspired fantastic creatures too. In September 1861, whilst Kingsley was writing *The Water Babies* and before Alice had fallen down her hole, 2 years after Darwin published *On the Origin*, a 55-year-old highly successful marine painter and amateur geologist called Edward Cooke attended the British Association for the Advancement of Science meeting in Manchester. It was the largest of the 75 annual gatherings which the Association held in the nineteenth century; in that year 3138 people attended. Some of these extra people had undoubtedly come for promise of more Darwin drama after the exchange of insults that enlivened the Oxford meeting the previous year. For a week Cooke dined late after sessions with fellow delegates, including his friend the zoologist Richard Owen, talking about Darwin's new book and its implications, discussing apes, archetypes and origins. A few days later, exhausted, he travelled to a seaside inn at Blue Anchor in Somerset, in order to do a few days sketching of fossils and cliffs. He found he was the only guest at the inn, but, as he began sketching he found himself drawing not fossils but a series of cartoon-like creatures, the first being the 'Ammonite with the Crested Cockatoo'. Over the next few days he drew more than a hundred hybrid monstrous creatures, endless forms, grotesque and curious com-binations of bones, shell, feather, tusks, scales and human limbs. He called them his 'Darwin Animals' (Figure 3.3).

Eight years later, Cooke published 50 or so of these drawings as a book aimed at the Christmas market entitled *Entwicklungsgeschichte*, the title,

FIGURE 3.2 Tenniel's drawing of Alice's encounter with the caterpillar in *Alice's Adventures in Wonderland*. The caterpillar asks Alice: Who are you?

the German word for evolution or history development, was translated later as *Grotesque Animals*. Each plate had a Latin motto pointing a moral. On the title page Cooke inscribed the lines:

> These oddities, from fancy drawn
> May surely raise the question
> Will Darwin say – by chance they're formed
> Or Natural Selection?

Cooke's monsters retain the structure of a human shape, but the bodies are made up of multiple cross-species animal body parts; a large proportion of these come from aquatic creatures. Most of Cooke's settings are shore lines, the site of the amphibian, the space in which terrestrial organisms emerged

FIGURE 3.3 'Darwin's Animals', Plate XIX from Edward Cooke, *Entwicklungsgeschichte*, 1872.

from earlier aquatic forms. Although these pictures are clearly part of a tradition of animal human grotesques made popular by the caricaturist J. J. Grandville in France earlier in the century in the *Metamorphosis du Jour* (1828–9) and before him Hieronymous Bosch and Pieter Breugel the Elder, there is something new emerging here, particularly when you look at these images alongside *Alice's Adventures in Wonderland* and *The Water Babies* and read them as a first generation response to Darwin's *On the Origin of Species*. Although Cooke's images are darker than Carroll's and Kingsley's, they come from a similar kind of reaction to the collapse of the border between the animal and the human.

A sense of dethronement

By the end of the nineteenth century a new pattern of literary response to Darwinism started to emerge. The idea of animal–human kinship began to turn into a series of nightmares of invasion and of degeneration as the beast ancestor, whether imagined as ape or sea creature, came to be represented over and over again in late nineteenth-century novels as outsized, threatening and invasive. We need think only of the Martians of *War of the Worlds* (1897), the Morlocks of *The Time Machine* (1895), the violent animal–human hybrids of *The Island of Dr Moreau* (1896), the shape-shifting vampires of *Dracula* (1897) and Hyde in *Dr Jekyll and Mr Hyde* (1886). The beast within can't, it seems, be contained any longer. It bursts out of the respectable Jekyll in the form of Hyde. It must be driven out. It must be exorcised over and over again.

We might borrow a phrase from H. G. Wells to describe this new anxiety. He called it 'a sense of dethronement'. H. G. Wells, founder of modern science fiction, was its greatest articulator. He trained as a zoologist at the Normal School of Science in South Kensington under T. H. Huxley and he knew Darwin's work well. By the time he began to write fiction, he had a first class honours degree in zoology and a second class honours in geology. When he came to writing *War of the Worlds* in 1897, a novel in which Martians invade the Earth and start to bring down all of its webs and networks starting with the railway system, the narrator describes an intense feeling of oppression:

> I felt the first inkling of a thing that presently grew quite clear in my mind, that oppressed me for many days, a sense of dethronement, a persuasion that I was no longer a master, but an animal among the animals, under the Martian heel.
>
> (Wells, 1897, p. 154)

Wells' Martians are outsized mechanised octopuses with legs. Except for the apelike Morlocks in *The Time Machine*, Wells' monsters are usually creatures of slime and tentacles rather than of fur or feathers. Even the Eloi are 'the half-bleached colour of the worms and things one sees preserved in spirit in a zoological museum' (Wells, 1895, p. 49). Perhaps the fear of being kin to these creatures is more visceral; I suspect it produces a different, perhaps even more profound, kind of uncanny.

Wells describes an intense feeling of nausea over and over again in his work. It is almost always directly related to this sense of dethronement he describes (the realisation of being an 'animal among animals') and it is almost always brought about by an encounter with something tentacled.

In *The Time Machine* (1895), for instance, the time traveller reaches the desolate shore of the end of the world and finds the very last living creature in it:

> A horror of this great darkness came on me. The cold, that smote
> to my marrow, and the pain I felt in breathing, overcame me. I shivered,
> and a deadly nausea seized me. Then like a red-hot bow in the sky
> appeared the edge of the sun. I got off the machine to recover myself.
> I felt giddy and incapable of facing the return journey. As I stood sick
> and confused I saw again the moving thing upon the shoal – there was
> no mistake now that it was a moving thing – against the red water of the
> sea. It was a round thing, the size of a football perhaps, or, it may be,
> bigger, and tentacles trailed down from it; it seemed black against the
> weltering blood-red water, and it was hopping fitfully about. Then
> I felt I was fainting. But a terrible dread of lying helpless in that
> remote and awful twilight sustained me while I clambered upon
> the saddle.
>
> (Wells, 1895, p. 78)

Wells, trained in zoology, composes his monsters, his Darwinian animal ancestors, from the body parts of sea creatures, not apes, and in this end-of-the-world encounter, he suggests to us that this is the primeval past and the primal form to which we will – or might – return.

Wells' work was in many ways an expression of a generation over-whelmed by the sense of dethronement he described. Britain no longer held its place in the world unchallenged. The period from 1870 to 1914 in Britain was a period of prolific invasion literature which was a sign of precisely this sense of dethronement, not only of a species but also of a nation and an empire. And Wells' invaders represent only a small part of the invading monsters of the *fin de siècle*. In Stevenson's *The Strange Case of Dr Jekyll and Mr Hyde*, the beast inside Jekyll ('my devil had long been caged, he came out roaring') is apelike: small, stunted and horrifying. For everyone who observes him he defies description. In Bram Stoker's *Dracula*, the vampire is an invading degenerate who can shapeshift, turning himself into a bat, rats, a dog, a lizard, wolves – he can cross the boundaries of time and of species. By this point in literary history, the fear of degeneration is of course uppermost in these stories – the fear that if human species can progress, then a reverse narrative might be possible too. So these narratives enact a series of expulsions. Dracula must be driven out. The Martians must be driven out. Hyde must be driven out. Animal–human kinship, it seems, is not to be borne.

By the early twentieth century, after the human slaughter of World War I and World War II, after Freud and Marx, after the horrors of the Holocaust, there was a marked, and perhaps inevitable, shift towards a greater accept-ance of the destructive and conflictual relationships at the heart of the natural world and of animal–human kinship, an acceptance of the dark forces at work within nature and within man, an acceptance of the war of nature. From the opening of Franz Kafka's *The Metamorphosis* (1915), in which travelling salesman Gregor Samsa wakes up as a giant insect, to the violent scenes of sexual competition and mastery in D. H. Lawrence's novels or the predatory animals and birds celebrated in the poetry of Ted Hughes or Tom Gunn or Dylan Thomas, it is possible to see that the argument for man's essentially animal nature had been won.

Entanglement

And what would Darwin have made of all of this? What would he have made of Lawrence and Hardy and Kafka? Would he have owned this way of seeing, this bleak refusal of happy endings, these tentacled nightmares of invasion and dethronement? Perhaps not. Almost certainly not. Yet the horror of famine and struggle and competition are there in Darwin's own writing. When Darwin strains his prose to new heights in order to persuade us to see something new, when he reaches towards a kind of sublime, it is almost always to try to convince us of the entanglement of terror and beauty. We need only recall that famous entangled bank passage at the end of *On the Origin of Species*:

> It is interesting to contemplate an entangled bank, clothed with many plants of many kinds, with birds singing on the bushes, with various insects flitting about, and with worms crawling through the damp earth, and to reflect that these elaborately constructed forms, so different from each other, and dependent on each other in so complex a manner, have all been produced by laws acting around us. [He gives the laws here.] Thus, from the war of nature, from famine and death, the most exalted

> object which we are capable of conceiving, namely, the production of the
> higher animals, directly follows. There is grandeur in this view of life, with
> its several powers, having been originally breathed into a few forms or
> into one; and that, whilst this planet has gone cycling on according to the
> fixed law of gravity, from so simple a beginning endless forms most
> beautiful and most wonderful have been, and are being, evolved.
>
> (Darwin, 1859, pp. 459–60)

In this passage Darwin reminds us that exalted objects emerge from
famine and death. He also stresses the endlessness and beauty of the
process in the exquisite rhythms of 'have been, and are being, evolved.'
This passage expresses a kind of Darwinian sublime. It is a form of prose
poetry. There is a second similar passage in the *Origin* that also fascinates
me. It too has a poetry in its rhythms and in its repetitions. He insists –
again – that we must see the entanglement of beauty and destruction:

> We behold the face of nature bright with gladness, we often see
> superabundance of food; we do not see, or we forget, that the birds which
> are idly singing round us mostly live on insects or seeds, and are thus
> constantly destroying life; or we forget how largely these songsters, or
> their eggs, or their nestlings, are destroyed by birds and beasts of prey;
> we do not always bear in mind, that though food may be now
> superabundant, it is not so at all seasons of each recurring year.
>
> (Darwin, 1859, p. 116)

*We behold but we do not see or we forget or we forget we do not always bear in
mind ... it is not so at all seasons ...* The language is elegiac. It is mournful,
clouded with sadness; it expresses a dark knowing. Even in the midst of
summer amidst the brightness and the abundance, he tells us, *we do not see
or we forget ... we forget* and then there is that beautiful echoing phrase: *we
do not always bear in mind.* We do not always bear. We cannot always bear.

There's a line very like it in George Eliot's *Middlemarch* where Eliot
ponders the question of how much we can bear to see or hear when she
describes the misery of Dorothea Brooke's early married life to the
dreadful Casaubon:

> That element of tragedy which lies in the very fact of frequency, has
> not yet wrought itself into the coarse emotion of mankind; and perhaps
> our frames could hardly bear much of it. If we had a keen vision and
> feeling of all ordinary human life, it would be like hearing the grass grow

and the squirrel's heart beat, and we should die of that roar which lies on the other side of silence. As it is, the quickest of us walk about well wadded with stupidity.

(Eliot, 1872, p. 194)

And it is this shadowing of the enchantment with horror or darkness that Darwin would have owned I think – that balancing of delight and terror that begins and ends in the beautifully tangled sentences of the *Origin*. This is the chiaroscuro of Darwinism. Darwin wanted us to see that and not to forget it. He wanted us to be able to bear it.

Of course, not all of the first generation of novelists who read *On the Origin of Species by Natural Selection* saw animal–human mutations or monsters. George Eliot saw webs. George Eliot saw webs *before* she read Darwin, as Gillian Beer describes in her marvellous book, *Darwin's Plots* (1983). George Eliot (the pseudonym for novelist and journalist Mary Ann Evans) was interested in human interconnections and in the nineteenth century, of course, it was Thomas Hardy and George Eliot whose fiction, complex and entangled, reveals something of the knowledge that *On the Origin of Species* shares – that society itself is a great entangled bank, a web of affinities, and that everyone, to use Darwin's phrase, 'is netted together'; a change in one place makes a change in another. George Eliot didn't get her sense of the web of affinities from Darwin but from her observations of her neighbours. She saw herself as a belated historian whose task was to look at the web of human society as she declared in the opening of Chapter 15 of *Middlemarch*:

> We belated historians must not linger over his example [Fielding's] ...
> I at least have so much to do in unravelling certain human lots, and
> seeing how they are woven and interwoven, that all the light I can
> command must be concentrated on this particular web, and not dispersed
> over that tempting range of relevancies called the universe.

Darwin's description of this complex, interdependent world, fruitful, diverse, woven together, in which every organism depends upon every other organism for its survival, and competes with it at the same time for survival, is something that George Eliot had already seen, something she already knew. She was already attending to this particular web of human relations and origins.

This sense of complex entanglement is evident in the work of many contemporary novelists too. In A. S. Byatt's (1990) beautifully entangled plot of *Possession*, which is shot through with references to Darwin and to entangled banks, Byatt shows us that past and present cannot be separated out – the business of the novel, she tells us, is to remind us of the relations between the living and the dead, between past and present. *Possession* celebrates shapeshifting, continual change, flux and entanglement. Order and chaos exist in mutual tension. It is a reflection of the benign version of the entangled bank – nature is plentiful, inter-connected, full of interest, sublime. We move from entanglement as chaos to entanglement as relatedness, entanglement as a web, a network. And it has a happy ending.

And there is Ian McEwan's (2005) *Saturday* which is also a novel shot through with Darwinian references, mostly missed by the reviewers. Ian McEwan's work as a short story writer and as a novelist has been a continual dialogue with Darwin and Darwinian webs. The whole novel of *Saturday* is a tangled bank. It encompasses a single day in the life of a successful neuroscientist. The book opens with Henry Perowne waking from his marital bed, his head full of Darwinian echoes from reading the last paragraphs of *On the Origin of Species* the night before. With one phrase from *Origin* resonating in his mind – *there is grandeur in this view of things* – he watches a plane on fire making an arc across the night sky. The day is shadowed by the threat of violence – on the edge of Perowne's bright and tangled bank the television set constantly reminds him of the terrorist threats and the violence in the streets.

As the day opens up, Perowne sees webs everywhere in the neural pathways of the brain, in the tangled roads of great cities. But he struggles to make sense of it. Perowne's daughter Daisy, a poet, gives her father books to read. It is Daisy who has given her father a biography of Darwin to help him understand. In the early evening of this particular Saturday, a day of homecomings, the comfort and beauty of Perowne's home is broken by the eruption of a violent and simian-featured boy called Baxter and his gang who threaten to kill his wife and rape his pregnant daughter. (In many ways, of course, *Saturday* is a rewriting of Stevenson's *The Strange Case of Dr Jekyll and Mr Hyde*.) In a strange scene, Daisy defuses the violence by reading a poem when the invaders taunt her to do so, a poem which seems to have particular importance to McEwan: Matthew Arnold's 'Dover Beach'. In the recitation of this poem, the tangled bank is complete, the boy, who has echoes of both Frankenstein's monster and Jekyll's Hyde, is overcome, and order is restored. In the final pages of the book, Perowne operates on the brain of Baxter, recalling the symbiotic, mutually repellent and mutually dependent relationship between scientist and monster in *Frankenstein*. As he looks into the network of neural pathways there, he marvels at the complexity of what he sees and returns to thinking about that phrase of Darwin's – *'there is grandeur in this view of life'*. Evolution, McEwan reminds us, may have created civilisation, morality, cities: it has also created the violence in Baxter's brain that threatens order.

By the second half of the twentieth century, then, there are fewer Darwinian monsters about in British novels. There is a return to some kind of beauty and enchantment in the Darwinian vision – a return to a preoccupation with webs and networks and what we would call the entangled bank not of human–animal kinship but of mutual dependency, migration and adaptation and a fascination too with chance encounters and collisions. And that, of course, is not surprising because, by the twenty-first century, webs, mutability and interdependence have become the dominant modes of our thinking about the world: communication networks, the World Wide Web and a global economy.

The poetics of the commonplace

Darwin claimed in his autobiography that in his latter years he lost his taste for poetry. He expressed this loss with regret:

> I have said that in one respect my mind has changed during the last twenty or thirty years. Up to the age of thirty, or beyond it, poetry of many kinds, such as the works of Milton, Gray, Byron, Wordsworth, Coleridge, and Shelley, gave me great pleasure, and even as a schoolboy I took intense delight in Shakespeare, especially in the historical plays. I have also said that formerly pictures gave me considerable, and music very great delight. But now for many years I cannot endure to read a line of poetry: I have tried lately to read Shakespeare, and found it so intolerably dull that it nauseated me.

> I have also almost lost any taste for pictures or music ... My mind
> seems to have become a kind of machine for grinding general laws
> out of large collections of facts ...
>
> (de Beer, 1974, pp. 83–4)

But poets have recognised the poet in Darwin. They have recognised in particular the way he moves through the commonplace, the ordinary, the overlooked, to the sublime. In 1964, Anne Stevenson wrote to the American poet Elizabeth Bishop about surrealism. She was writing a book about Bishop and she wanted to make a connection between Bishop's poetry and surrealism. 'No,' Bishop replied, 'I'm not like the surrealists, I'm like Darwin':

> Reading Darwin, one admires the beautiful and solid case being built
> up out of his endless heroic observations, almost unconscious or
> automatic – and then comes a sudden relaxation, a forgetful phrase,
> and one feels the strangeness of his undertaking, sees the lonely young
> man, his eyes fixed on facts and minute details, sinking or sliding
> giddily off into the unknown.
>
> (Letter 8, 20th Jan; Bishop, 1964)

What Elizabeth Bishop recognised in Darwin's writing is what I would call the poetics of the commonplace. I imagine Bishop might have put Darwin straight when he expressed his sadness about losing his taste for poetry and lamented that he had become a mere grinder of facts. The grinding of facts is where poetry begins, she might have said. The sublime is reached through the commonplace, through the slow accretion of facts. Look at what you do, she might have said, look at the way in which you make us slide or sink with you, through the facts, giddily into the unknown.

Darwin startlingly once told his son Francis that he might write a poem about *Drosera* – the sundew plant on which he was working in preparation for his book on insectivorous plants. The *Drosera*, or sundew, so named because of the sticky droplets it produces to catch insects and which look like dewdrops, is a tiny carnivorous plant. There are 170 different species and they grow on all continents except Antarctica. Darwin found one of these plants in 1860 and brought it home to do experiments on it. It fascinated him. Unless there's a piece of paper still unearthed by the people in the Darwin-Online Project or the

Darwin Correspondence Project, Darwin did not write that poem about the sundew.

But Amy Clampitt did. In 1978, she published a poem called 'The Sun underfoot amongst Sundews' in *The New Yorker*; it was her first published poem. She was 58. In 1983, at the age of 63, she published her first full length collection, *The Kingfisher*, and then five more collections before she died in 1994 (Clampitt, 1998). These poems are exquisite expressions of what we might call the Darwinian sublime. In this extraordinary poem she asks us to imagine stepping into a bog full of sundews, a metaphor for our lives; she reminds us that we will be swallowed up, that we will not *get out of here*. But there is so much to see, she says, so much light, so much of the sublime. If we look properly, she says, once we begin to see the sublime beauty here in this Darwinian underworld, we will begin to *fall upward*.

> An ingenuity too astonishing
> to be quite fortuitous is
> this bog full of sundews, sphagnum-
> lined and shaped like a teacup.
>   A step
> down and you're into it; a
> wilderness swallows you up:
> ankle-, then knee-, then midriff-
> to-shoulder-deep in wetfooted
> understory, an overhead
> spruce-tamarack horizon hinting
> you'll never get out of here.
>   But the sun
> among the sundews, down there,
> is so bright, an underfoot
> webwork of carnivorous rubies,
> a star-swarm thick as the gnats
> they're set to catch, delectable
> double-faced cockleburs, each
> hair-tip a sticky mirror
> afire with sunlight, a million
> of them and again a million,
> each mirror a trap set to
> unhand believing,

            that either
a First Cause said once, 'Let there
be sundews,' and there were, or they've
made their way here unaided
other than by that backhand, round-
about refusal to assume responsibility
known as Natural Selection.
               But the sun
underfoot is so dazzling
down there among the sundews,
there is so much light
in that cup that, looking,
you start to fall upward.

4 Darwin and human society

PAUL SEABRIGHT[1]

Charles Darwin has long been associated in the mind of the educated public with a vision of competition as fierce, unforgiving and absolutely omnipresent. That vision has haunted the late 19th and 20th centuries, a period when both economic and military rivalry reached new heights of ambition and brutality. It has inspired visual imagery and social meta-phor, viewing competition as almost a disease, a predicament of the mass in human society, pitting us, each of us, against all the others.

T. S. Eliot's *The Wasteland*, published four years after the end of the First World War, gave bleak expression to this vision:

> A crowd flowed over London Bridge, so many,
> I had not thought death had undone so many.
> Sighs, short and infrequent, were exhaled,
> And each man fixed his eyes before his feet.

If you enter those words into Google™ and perform an image search, you will find that *The Wasteland* has stimulated the photographic imagination of people all over the world. Figure 4.1 shows an early photograph of the crowd flowing over London Bridge in around 1904. The internet will show you others taken more recently on London Bridge, depicting alienated, anxious individuals on their way to work – but it's not just from London that you find such photographs. Google™ brings us images of competition in the crowd, of alienation from our fellow human beings, from places as far apart as Osaka and St Petersburg. Figure 4.2 shows an extraordinary long-exposure

Darwin, eds. William Brown and Andrew C. Fabian. Published by Cambridge University Press. © Darwin College 2010.

FIGURE 4.1 London Bridge, around 1904. Copyright: Corbis.

photograph by Alexey Titarenko from St Petersburg, in which individuals have disappeared into the mass, streaming who knows where, to some rendezvous with an indifferent destiny.

If we can escape being intoxicated by the poetry, and look for a moment at the world around us as we meet it day by day, almost everything about this picture of competition and its relation to our social life is inaccurate. Think what happens when you leave your house in the morning. Unless you live alone you will be leaving a household full of people with whom you need to cooperate. If you take public transport to work, the bus or train driver is someone on whose competent fulfilment of their function you absolutely depend; this is not a competitor, this is a collaborator. You may buy your newspaper from somebody who has thoughtfully made it available for you to buy; this is a collaborator not a competitor. You then arrive at your place of work. Economists would recognise that the test of whether somebody is a collaborator is by and large whether you are

FIGURE 4.2 City of shadows: St Petersburg, 1990s. Copyright: Alexey Titarenko.

pleased when they do a good job. A competitor, by contrast, is someone whose good performance of their job is a threat to you. How many of you arrive at your factory, office or shop and are pleased to discover that the other people in your workplace have done a bad job?

In short, the picture that I just painted for you, allegedly the one bequeathed to us by Charles Darwin, a picture of unforgiving competition everywhere, makes no sense of the ties of collaboration that we have with large numbers of the people we meet in our professional and social lives.

Figure 4.3 also shows a photograph of a crowd flowing over London Bridge. At first sight it might strike the viewer also as depicting a mass of alienated individuals, each man fixing his eyes before his feet. But if you look more closely it is actually a group of strikers crossing London Bridge – this is from the London dock strike of 1911 – united on a common purpose. Pictures can be read many ways. If we have uncritically absorbed the notion of competition as everywhere around us, we may interpret photographs like these as corroborating our view, when in fact

FIGURE 4.3 Strikers crossing London Bridge, 1911. Copyright: Library of Congress.

the truth is different. The story I want to recount here is about how the difference of that truth was evident to Charles Darwin. If so many of his contemporaries failed to see it, that is because they were not looking where he was looking.

Still, let me start with a puzzle. Suppose everything I've said is true, that cooperation is everywhere, and that on our daily ride into work we meet far more of our collaborators than our competitors. . . why does it all feel so stressful?

Questions about the role of stress in social life are not ones economists are used to thinking about. But they would not faze a primatologist. Primatologists know that human societies are highly structured primate groups. Like almost all primate societies, our own is characterised by a highly complex interaction of cooperation within groups and competition between groups, and it is a process that can be highly stressful. I shall return to the subject of stress in a moment, but let me first summarise what I shall cover in this short chapter. First, I shall discuss some of the things we have in common with our primate cousins, as well as some of the things that set us apart from them. Secondly, I shall introduce the subject of sexual selection, which was central to Darwin's understanding of why human societies had evolved. In particular, it was his means of explaining why there was such apparent diversity among the various human populations when, as Darwin firmly believed, all living human beings could trace their descent from a common ancestor who was already fully human. Finally, I shall ask how much Darwin's own views on racial and gender differences in human populations stand up in the light of our modern understanding.

Cooperation and competition in primate societies

Let me begin by showing how clearly Darwin repudiated the view that cooperation is marginal to human social life. Here is a passage from *The Descent of Man*:

> When two tribes of primeval man, living in the same country, came into competition, if (other circumstances being equal) the one tribe included a great number of courageous, sympathetic and faithful members, who were always ready to warn each other of danger, to aid and defend each other, this tribe would succeed better and conquer the other. Selfish and contentious people will not cohere and without coherence nothing can be effected. A tribe rich in the above qualities would spread and be victorious over other tribes. . . thus the social and moral qualities would tend slowly to advance and be diffused throughout the world.
>
> (Darwin, 1871, part I, pp. 162–3)

Not only did Darwin clearly believe that human beings were capable of the social and moral qualities, but he was convinced that natural selection would encourage these qualities.

Darwin also clearly believed in the heritability of behaviour and not just of physiology. There is no question about this: anyone who doubts it should read the wonderful chapter on bees in *The Origin of Species*. But though he thought behaviour could be inherited, he also believed that tiny differences in predispositions to behave in certain ways could have remarkable effects upon the structure of social behaviour in the mass.

His discussion of the architecture of beehives shows clear awareness that the effects of the interaction of many individuals may not be ones that are themselves coded or hardwired in the particular instincts that make them possible. Darwin also believed very passionately in the major influence of environmental factors on social behaviour; nobody who reads his discussions of the human races in *The Descent of Man* can realistically doubt that. While in the late 19th century and for much of the 20th century, explanations of behaviour in terms of heredity and of environment were widely perceived as competing explanations, it is now possible to see these influences as fundamentally complementary. Linguistics, for example, has made it clear that even if there is an astonishing variety of languages that are spoken across the world, they represent but a small subset of all the languages that could theoretically be spoken (Pinker, 1994). There is important work to do to explain why languages differ from each other, just as there is important work to do to explain why languages are similar to each other. It's also clear that without some idea of the hardwired factors that predispose us to learn in certain ways, we could not understand why languages have the wonderfully diverse structure that they have. In other words, we need to understand inheritance to understand how the environment affects us in the way it does. This has become true now for a number of other areas of social life and not just for linguistics.

The fundamental principle to bear in mind in what follows is encapsulated in the wonderful phrase that concludes *The Descent of Man*: 'Man still bears in his bodily frame the indelible stamp of his lowly origin' (Darwin, 1871, part II, p. 405). Those tempted to read it as a gloomy statement about our animal nature and the way it pollutes our social behaviour might compare it with the closing paragraph of the *Origin*, where he writes:

> There is grandeur in this view of life, with its several powers, having been originally breathed into a few forms or into one; and that, whilst this planet has gone cycling on according to the fixed law of gravity, from so simple a beginning endless forms most beautiful and most wonderful have been, and are being, evolved.
>
> (Darwin, 1859, pp. 459–60)

For Darwin, our humble origins were a matter of pride rather than reproach, and this was no less true for the fact that our origins left their mark. So let me now ask what we learn from our kinship with primates. We are, as Darwin was very much aware, group-living primates for whom major determinants of fitness include our ability to cooperate with others in our group, the social intelligence to work out who else can be trusted to cooperate with us, and the ability to work upon the other members of our group to induce them to work with us and not against us.[2]

Within primate groups there is indeed frequent competition between individuals over basic access to economic resources. But competition also takes place in other dimensions, notably between groups and coalitions of individuals over access to resources, and between individuals over the right of access to powerful groups. The tension between those forms of competition is at the heart of primate social life. At any moment you are seeking to cooperate with some of the other members of your group, but at the same time you are intensely anxious about which groups you may be allowed to join. Your fitness is going to depend not just on what you do but on what you can induce the other members of your group to do with you and for you.

This means that in the psychology of being a primate there is a continual tension between our talents for cooperation, which are many and sophisticated, and our predispositions towards competition. This sometimes takes a very violent form, and it's almost always stressful, because very small status differences can have very large fitness consequences. Anybody wondering whether primates are responsible for the recent financial meltdown of the world economy cannot but have been struck by the way in which very small apparent differences in talent have led to very large differences in remuneration; our economic structures faithfully reflect our primate past.

In dominance hierarchies in primate societies, there is an association between low status and various characteristics that are usually good predictors of stress, such as a lack of autonomy and social control, and a high degree of unpredictability in outcomes. Studies of human societies suggest something very similar: a longitudinal study of British civil servants by Michael Marmot and others shows that stress-related illnesses such as cardiac disease are not associated with high status jobs as is popularly believed, but have instead a strong negative correlation with rank (Marmot, 2004). So the lower down the hierarchy you are, the more likely you are to have somebody pushing you around, telling you what to do with your life. This is a very stressful thing to happen to you, and is associated over time with significantly greater risk of stress-related illnesses such as cardiac disease.

It was widely believed until recently that in non-human primate groups exactly the same thing was true, but when it was first possible to measure cortisol levels of primate groups in the wild, some slightly surprising results emerged. An interesting paper by Muller and Wrangham showed that cortisol levels were *positively* correlated with dominance among chimpanzees in the wild (Muller and Wrangham, 2004). The authors suggest that this is because dominant chimpanzees in the wild spend a great deal of energy maintaining their dominance – more so, certainly, than senior civil servants have to do. It's a tough and uncertain life up at the top in a chimpanzee group, and your cortisol levels are elevated because of the continual uncertainty about how long you're going to stay on top and the continual efforts you have to make to ensure you stay there. Frans de Waal's wonderful book, *Chimpanzee Politics*, describes such a life in gripping detail (de Waal, 1982).

So primate existence is pretty stressful either way. You get stress if you win and stress if you lose. The explanation seems to be precisely that very small differences in behaviour result in very big differences in outcomes. It's not surprising therefore that even though we're surrounded by potential collaborators, we live in continual fear that our potential collaborators will decide to collaborate not with us but with somebody else.

If this is a predicament we share with other primates, what makes us different? The answer is that many things do, most of which I shall not deal with further here despite their obvious importance – things such as language and the elaborate use of culture. Instead I want to focus on the particular brains and bodies and social organisation that distinguish us from other primates.

The first important difference is that we are very environmentally flexible, living in a much wider range of environments than other primates, and as a result we have very high returns on learning. This is not true just of people in modern societies: it is true of hunter–gatherers too. Work by Hillard Kaplan and a number of collaborators

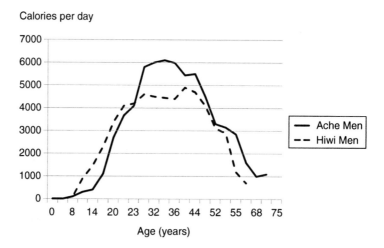

FIGURE 4.4 Daily energy production by male Ache and Hiwi hunter–gatherers as a function of age. Source: adapted from Kaplan *et al.*, 2000.

has shown that even in hunter–gatherer groups, the returns on experience and human capital continue to be high until people are in their mid 40s. Figure 4.4, drawn from their work, shows the daily calorie production of men in two hunter–gatherer groups plotted against the age of the men concerned (Kaplan *et al.*, 2000). The productivity of male hunters peaks around 35 in one group and in the early 40s in the other, although upper body strength for males is greater when they are in their early 20s than it will ever be again. In other words, hunting is an activity in which the returns are not so much down to physical strength as to cunning, experience and knowledge. This also tells us something about how important it was for human beings to have a reasonably long lifespan, because there is no point in investing in the ability to hunt in sophisticated ways if you're not going to be around long enough to reap the returns on that investment.

Among the pressures that led to the evolution of human beings with large enough brains to be able to learn, the reward of hunting was obviously one of the leading factors. But that adaptive pressure came with a cost. Creatures with very large brains have to develop them either in the womb during gestation or after birth. In early human evolution

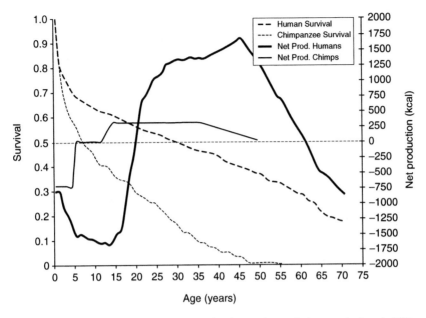

FIGURE 4.5 Net calorie production and cumulative survival probability
as a function of age for chimpanzees and humans. Source: Kaplan *et al.*, 2009.

our upright stance set a natural limit to the breadth of the female pelvis,
which ensured that human babies had to be born systematically premature. So they became dependent on their parents for a long period after
birth; it was only because there were high enough returns to being large-
brained later in life that it was possible to incur the very high costs of
being so dependent early in life.

Figure 4.5, also from work by Kaplan and collaborators, shows the
remarkable contrast between the profile of the energy costs and benefits
of being a chimpanzee over life, and the equivalent profile for human
beings (Kaplan *et al.*, 2009). Until the age of about five, the chimpanzee is
dependent on adults. But after that it becomes reasonably self-sufficient
until the point that it's caring for young in its turn, but it's not producing
a very large surplus. In human groups, by contrast, young individuals
have a net calorie deficit until nearly the age of 20. These data are
averages for several groups and the story is not the same in every group,
but essentially young humans are dependent for a very long time.

Many people find these data surprising because they see adolescent boys grow up and become rather strong; thinking that strength matters in hunting they assume that by the age of 12 or 13 a young hunter–gatherer male would probably be paying for his keep. It turns out not to happen, because the young hunter–gatherer male is eating a great deal. So it's really not until age 20 that the hunter–gatherer male starts to pay back, and then he has to pay back a lot and he has to pay back for a long time. So that's a clue as to how the cooperation across generations in hunter–gatherer groups has to be dramatically more sophisticated than it is in chimpanzees.

Figure 4.5 also shows the cumulative survival of chimpanzees and humans over time. There is very high mortality among the young but the curve is steep everywhere. For human hunter–gatherers, there is a very steep part of the curve at the beginning because of infant and child mortality, but then it takes a much more serene downwards slope. So the greater expected lifespan is what underpins the greater productivity of the adult male in humans. This makes us quite different from all other primate groups, and means that we are collaborating across generations in a way previously unknown in Nature.

The result has been to raise dramatically the returns to both pair bonding and three-generational resource interdependence. Figure 4.6 shows the net transfers in kilocalories per day between different generations of hunter–gatherer groups. Transfers from fathers to children, for example, are very high in the mid 40s (Kaplan *et al.*, 2009). Interestingly, transfers from grandparents to grandchildren are important as well. These resource transfers are an important part of the overall ecology of a hunter–gatherer group, and they are part of what made us distinctively human.

What does all that mean? It means that we've been used, since well before agriculture and civilisation, to elaborate networks of cooperation, and in particular to cooperation involving long-term trust and deferred gratification. While there is some evidence of trade within chimpanzee groups (for instance, male chimpanzees who have been successful in a hunt use meat to trade for sex with female chimpanzees), much of the trade seems to be essentially simultaneous (Stanford, 1999). A very recent study by Cristina Gomes and Christophe Boesch has for

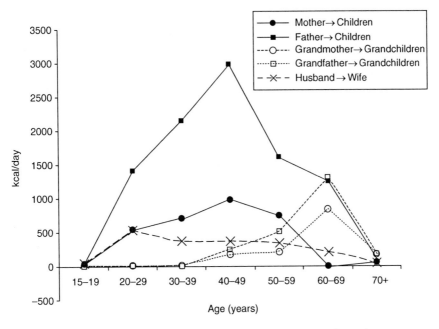

FIGURE 4.6 Net resource flows as a function of age for male
hunter–gatherers. Source: Kaplan *et al.*, 2009.

the first time uncovered evidence that meat-trading among chimpanzees
leads to bonds that persist over time (Gomes and Boesch, 2009), but
there remains a striking contrast with human societies. Compared to
those of other primate species, human societies are based on a much more
elaborate capacity for deferred gratification in the service of complex
social purposes.

We also interact outside close kin. While that's obviously true in
modern sophisticated economies, even in hunter–gatherer groups there
are complex types of operation, which don't take place just between close
relatives. This emphasis on cooperation makes sense of a remarkable
puzzle about human societies which is the subject of a fascinating book
by Christopher Boehm, called *Hierarchy in the Forest* (Boehm, 1999). Lest
you think the story I've been telling you so far is one about primate
genes determining us to live in a primate way, Boehm's work depicts a
primate nature manifesting itself in social behaviour that has proved

highly responsive to variations in our environment over the ages. In particular he is intrigued by the puzzle that most primate species other than our own are very hierarchical, and modern humans are fairly hierarchical, but almost all of the evidence that we've been able to gather about the lifestyle of hunter–gatherer groups (the stage of our history lying between our ape past and our civilised present) has suggested that they were remarkably egalitarian. How could this have happened?

Boehm's answer appeals not to some egalitarian instinct that we have now lost but rather to a countervailing force offsetting our very strong tendency to compete (Boehm, 1999). The strong sense of status that we share with all other primate species meets in human beings a tendency for coalitions of the weak to form to counterbalance the overweening behaviour of the strong. In modern societies the strong have ways to fight back, but in many hunter–gatherer societies the strong (the people who are a bit better at hunting, a bit more manipulative, a bit more able to marshal the resources of the group) are nevertheless aware that if they abuse any of the privileges this gives them, the weak may unite against them. Of course this is possible only so long as coalitions of the weak are able to marshal the resources that make them able to take on the strong. One of the reasons why modern societies are not as egalitarian as hunter–gatherer societies is that they have been able to build surpluses through agriculture and storable production that have led to concentrations of power – the strong can hire the services of mercenaries to ensure that the weak cannot effectively fight back. But in hunter–gatherer societies you can't do that. Hunter–gatherer strong men are too dependent upon the support and the collaboration of those around them to be able to exert dominance purely through fear.

As I mentioned, Darwin himself was very aware that the cooperation we see in human societies, both today and in the past, has been based upon a very strong degree of competition between groups. This competition has in the past been extraordinarily violent. To give a sense of the magnitude of the changes we have seen, in non-state societies (both existent societies that do not have a state and in such societies of the past), estimates of the proportion of deaths due to violence range between around 15% and 30% of all deaths. That is a very, very big number by modern standards. Lawrence Keeley, for instance, gives a range of different estimates of the proportion of deaths due to violence, both in existing non-state societies and in historical societies, based on estimates from archaeological remains (Keeley, 1996). Samuel Bowles has undertaken a more recent review of the evidence and estimates a mean mortality from violence of 14% among foragers (which is likely to have been less than that among early agriculturists) (Bowles, 2009). It's difficult to estimate the proportion of people who have died from violence because a lot of violence doesn't leave marks on skeletons. We also know that skeletons that are found that have clearly died from violence may not be representative samples of all the people who have died, so there is much controversy surrounding these estimates, which is also reflected in their high variance. But though the variance is very high, their mean is also extremely high.

The mortality level in the world today is tiny by comparison. Today the rates of violent death are around 1.3% of all deaths. That includes war, urban violence, everything; remarkably, it's even less than the rate of suicide. It's only a little over half of the rate of death from road accidents. As a random citizen of the world today you are more likely to die at your own hand than at the hand of somebody else – unless that person happens to be driving a car.[3]

So something rather remarkable has happened. The high mortality from violent competition between groups in the primate societies from which we evolved, and which is also extremely high in chimpanzees and in other primate societies, has fallen dramatically during the period since we adopted a state. Furthermore, the explanation is clearly cultural and not genetic. This is in no way inconsistent with the claim that our primate nature strongly influences our social behaviour. On the contrary, primates are animals that are particularly adapted to respond to the incentives in their environment. Richard Wrangham and Dale Peterson, for example, have written about primate violence, emphasising the extent to which violence exercised by chimpanzees is highly dependent upon context and opportunity (Wrangham and Peterson, 1996). Chimpanzees can be entirely peaceful with each other until an opportunity comes to be violent and get away with it. Human beings are the same.

Figure 4.7, drawn from the work of Manuel Eisner, provides evidence about the way in which homicide rates evolved in England from the

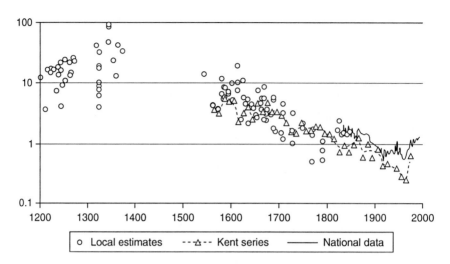

FIGURE 4.7 Eight centuries of English homicide rates. Annual homicides per 100 000 population. Source: Eisner, 2001 (Figure 4.1).

13th century to the 20th century, based on a large number of studies of village records. There is a very steady downward trend in the homicide rates over that period of time. No single factor was at work as far as we know. What happened was the gradual development of a set of institutions that simply made it less rewarding for potentially violent individuals to carry that violence out. The English evidence is entirely consistent with that from other European countries, as shown in Figure 4.8 from the same source (Eisner, 2001). Italy really was more violent in the Middle Ages (it wasn't just Shakespeare's imagination), but England, Italy, the Netherlands, Scandinavia, Germany and Switzerland all saw a steady decline in rates of homicide over that period. Once again, there was no magic bullet; this is best understood as a steady development of institutions that made it more profitable for individuals to settle their differences by non-violent means.

Sexual selection and the diversity of mankind

Darwin's theory of sexual selection is the key to understanding his ideas about the descent of man, for three main reasons. First of all, it was for Darwin the most convincing explanation for divergent evolution, which

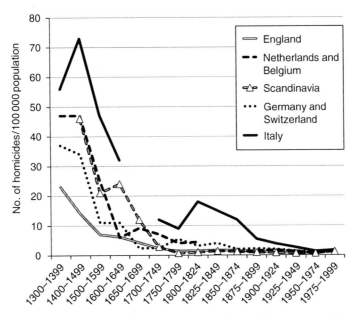

FIGURE 4.8 Seven centuries of European homicide rates. Annual homicides per 100 000 population. Source: plotted from data in Eisner, 2001 (Table 1).

is what occurs when two or more populations can come to diverge from each other very rapidly in their phenotypic characteristics. Adrian Desmond and James Moore have recently argued in *Darwin's Sacred Cause* that Darwin was strongly motivated by his detestation of the institution of slavery, which he had seen at close hand during the voyage of the *Beagle* (Desmond and Moore, 2009). He reacted in particular against those defenders of slavery who argued that the black races were a different species from the white races. Darwin hoped that if it could be shown that all the human races were descended from a single human ancestor, the case for mistreating the members of what he called the savage races would be much weaker. In order to do that he had to explain why the black races and the white races and the brown races looked so different when they were really so similar. Sexual selection came to be crucial to that explanation.

Secondly, for those who associate the notion of Darwinism with some kind of perfectionism or functionalism in the explanation of human behaviour, it's important also to realise that sexual selection shows clearly how evolution by natural selection can produce extraordinarily inefficient and repellent forms of behaviour. Darwin himself famously wrote 'what a book a devil's chaplain might write about the clumsy, wasteful, blundering, low and horribly cruel work of nature'.[4] For people who have seen how admiringly Darwin wrote about the beehive, and about some of the organs of the animal body, it can sometimes be a surprise that he was completely aware of the more rebarbative results of evolution by natural selection. Sexual selection, though not the only mechanism of natural selection that can explain this, nevertheless gives us perhaps the clearest view of how it can happen.

Thirdly, Darwin's view of sexual selection brings home vividly the tense interplay between cooperation and competition in social life.[5] Anybody who has ever been involved in sexual relations must have pondered the paradox that sexual reproduction is in some respects the most cooperative activity that any human beings can undertake, and yet the search to be accepted as a sexual partner can be one of the fiercest and most unforgiving forms of competition on the planet. That paradox was at the heart of Darwin's vision both of sexual selection and of competition in human society as a whole. It was also a moving reminder that nothing important in life is accomplished entirely alone.

Let me set out the main elements of sexual selection so as to show how the theory delivers on all of these three promises. Females are defined in Nature as the producers of the larger gametes among the two sexes and these are relatively scarce. Human females produce an egg a month, whereas males produce a thousand sperm a second. In the time it takes the female to produce that one egg, therefore, her male partner will have produced enough sperm to fertilise all the women of reproductive age in the whole world. Obviously the constraints on his ability to do so are not just logistical. They consist essentially of competition from rival males and the selectivity of females. One result of this asymmetry is that males essentially benefit more in fitness terms than females do from strategies which increase the quantity of their mating opportunities – either by defeating rival males or by overcoming female selectiveness. And the way they overcome female selectiveness is by offering or signalling the presence of the scarce resources that females need.

It might seem that the second strategy, persuading females to accept mating rather than beating up rival males, might lead to the interests of males and females being aligned. But in fact, as the study of many species in Nature makes clear, this is true only for the current mating. Their interests for subsequent matings may be very different and may affect how they approach the current one.[6]

It's worth noting that female selectiveness doesn't typically mean monogamy. In many species the females can be highly selective about their choice of mating partners without having to limit their choice just to one. Even in socially monogamous animals, including many birds, DNA studies have shown significant proportions of a female's eggs to have been fertilised by a male other than the female's partner.

The manifestations of sexual selection depend critically on the balance between force and persuasion in male reproductive strategies. Although some consequences of sexual selection appear to be common across species, like a greater established tendency for males in most species to take risks, most of them vary greatly from one species to another and even from one habitat to another. The reason for this, as game theorists would put it, is that the strategies of males and females in any one species are best responses to each other; the strategies are co-evolving in what may be an evolutionary arms race. And this makes divergent evolution a very common outcome.

Substantially divergent evolution between closely related populations is not necessarily something we should expect natural selection to deliver. In fact it is striking how often natural selection has produced convergent evolution – it has found functionally similar solutions to the same problem at multiple occasions throughout the history of life on our planet. Thus birds and bats have both evolved wings. Arctic and Antarctic fishes have both evolved a way of stopping their blood from freezing in the cold waters, by synthesising antifreeze proteins; but they do so using different proteins and with quite different genes that encode them. Anteaters in Australia and South America have developed long snouts, but they are not at all related (one is a marsupial) and the common ancestor almost certainly didn't have such a snout (Coyne, 2009, p. 92). The gene for lysozyme has evolved convergently in cellulose-digesting mammals (Ridley, 2004, p. 186). Garter snakes and clams have independently evolved a very similar mechanism giving them resistance to toxins in their prey.[7]

So natural selection against a reasonably static environment is often hitting on the same solution several times. Sexual selection, in contrast, is about co-evolving with somebody else, a mate or a rival. Therefore sexual selection is trying to hit a moving target, and two closely related populations can move the target in quite different directions. Again and again we see examples of divergent evolution between closely related species, which have similar habitats and similar histories but have hit on different solutions to the sexual selection problem.

This co-evolution is also the source of sexual conflict, which occurs even over when and how to mate. This is because the selectivity of females creates an adaptive advantage for strategies of male persistence, and that in turn favours the evolution of female selectivity even more. It's not just adolescent human males who can't understand why females aren't willing to have sex with anyone at any time. It's also true of elephant seals and water striders and tunnel web spiders and scorpions and bedbugs.[8] I don't have the space to give you the full soap opera, but elephant seal males are constantly forcing themselves on unwilling females. Water striders have evolved hooks on their antennae whose sole purpose is to hold the females down when the females are trying to escape. Tunnel web spiders anaesthetise their mates with a powerful toxin and mate while they are unconscious. Certain scorpions have evolved a kinder strategy: the toxin they use for hunting is too powerful, and would kill the female, so they have thoughtfully evolved a slightly less powerful drug that merely sends her to sleep. And the bedbug, *Cimex lectularius*, punctures the female's abdomen with a dagger, inserting the sperm directly into the body, because protective armour evolved around the female's vagina has become too difficult for the male to circumvent. No one contemplating this parade of bestial rape could remain convinced that behaviour that is adaptive under natural selection is always efficient, pleasant or even remotely optimal. That is crucial to understanding how Darwin saw evolution as applying to human societies.

When males compete for access to females by persuasion and seduction, we expect selection not just for physical strength but for display. In a large number of species, notably in many birds, this has led to the development of colourful plumage in males, and only in males. However, extensive research conducted in the challenging environment of the Cambridge May Ball season has established that at least one primate species, namely our own, has done things very differently. The colourful individuals on the dance floor that look like males are in fact the females, while the drab ones in black and white are actually the males (yes, really). As the example suggests, primates have done sexual selection a bit differently. In fact, human beings, like some other primates, seem to be displaying in both directions. The reasons for this are very interesting.

When males compete not just to impress females with their looks and talent, but also by cornering a scarce resource that females need, can this persuade females in turn to compete for access to that scarce resource? The answer is yes in principle, but what might that scarce resource be? It cannot just be the male sperm, however high its quality, because sperm is very abundant, so that any female who wants to can in theory have some. But when males become monogamous (to some extent) or when they contribute something genuinely scarce such as food, then female competition can become intense. The dance fly *Rhampomyia longicauda* illustrates this beautifully (Arnqvist and Rowe, 2005, pp. 74–7). Females swarm to mate. Males enter these swarms bearing small arthropods they have killed as gifts for the females, who feed on them while they mate. It becomes important to the females to capture a male who has economic resources, so the females compete for the males. They have evolved air sacs which are designed to look big and fertile, as though they are full of eggs, just as artificial breast enhancement in *Homo sapiens* aims to raise sexual desirability by mimicking the assets of the healthy and fertile female. But the dance fly's solution is cheaper than silicone; it only needs air. And it really works; the males go wild.

What does this mean for gender relations in apes and especially for humans? We apes have of course done things very differently from dance flies, but we seem to have learned a thing or two from them. Among our cousins the chimpanzees and bonobos, females enjoy a significant degree of independence and autonomy from males. They still face attempts to control them on the part of males, but they follow their own reproductive agenda, as the primatologist Craig Stanford has put it (Stanford, 1999).

This state of affairs had to change for human females, because they needed both protection and protein for the brains of their growing young. Both of these were scarce resources, and human females learned to compete for them, and competition exacted a price. The price was some loss of autonomy for human females, though this was limited by the need for females in hunter–gatherer groups to have enough freedom to gather effectively. The price became heavier when human beings settled down to farming, because farmers are able to confine their females much more effectively than hunter–gatherers can do. Now that economies are no longer based on agriculture, we may well wonder to what extent it is possible for gender relations to escape the apparent continual willingness of males to try to confine their females, and to be able to do so because of the dependence of females on the male supply of economic resources. While male desire to limit female autonomy does not seem to have changed much since prehistoric times, the complete economic dependence of females, which allowed that desire to be satisfied, no longer holds in advanced modern societies.

One last piece of relevant evidence about primates concerns differences in the way in which male and female chimpanzees form coalitions. Conflicts between female chimpanzees are much rarer than between male chimpanzees but they are much less often reconciled. A study in Arnhem Zoo by Frans de Waal, whose book *Chimpanzee Politics* I cited earlier, found that reconciliation occurred after around half of the conflicts between adult males but after less than one in five of the conflicts between the females. And when males intervene to try and break up a conflict between others, they are more likely to do so for strategic motives rather than in support of a friend. Males are more willing than females to reconcile with their enemies, and more willing to betray their friends (de Waal, 1989, pp. 48–57).

The result of this is that male coalitions are comparatively strategic and flexible, so that males who have been fighting can bury their differences to cooperate if a suitable foraging opportunity arises. Female coalitions are more loyal, stable and supportive, but less able to respond opportunistically to foraging opportunities. Similar findings have been reported in rhesus monkeys. In terms made famous by the sociologist Mark Granovetter, the males seem to have more weak ties and fewer strong ties in their networks than females do (Granovetter, 1973). I return to this below.

Darwin on sex and race

So what did Darwin himself make of all of this? There is some strong stuff in *The Descent of Man* about females exercising sexual choice, not a particularly popular topic in Victorian England. He wrote that 'in utterly barbarous tribes the women have more power in choosing, rejecting and tempting their lovers or of afterwards changing their husbands than might have been expected' (Darwin, 1871, part II, p. 373). It's possible that Darwin's rather revolutionary emphasis on the autonomy and the power of choice of females was partly responsible for the comparatively weak impact his theory of sexual selection had on the scientific community until well into the 20th century.

Darwin also wrote some things that look pretty odd to modern sensibilities. For instance, that:

> The chief distinction in the intellectual power of the two sexes is shown by man attaining to a higher eminence in whatever he takes up than a woman can attain.
>
> (Darwin, 1871, part II, p. 327)

And in a sentence which would certainly have denied him the presidency of Harvard nowadays, he wrote:

> It is indeed fortunate that the law of the equal transmission of characters to both sexes has commonly prevailed through the whole class of mammals, otherwise it is probable that man would have become as superior in mental endowment to woman as the peacock is in ornamental plumage to the peahen.
>
> (Darwin, 1871, part II, p. 329)

We need to remember that Darwin, though astonishingly prescient, was in many ways a child of his time. This is even more strongly marked when we look at what he wrote about race.

I have been persuaded by Desmond and Moore that the experience of slavery, which Darwin saw at first hand on the *Beagle*, was devastating for him, and led him to wish to emphasise the common character of the human races. In *The Expression of the Emotions in Man and Animals,*

Darwin not only asserts repeatedly that the different races of men express their emotions with remarkable uniformity throughout the world, but then draws an important conclusion from this. He says:

> All the chief expressions exhibited by man are the same throughout the world. This fact is interesting as it affords a new argument in favour of the several races being descended from a single parent stock which must have been almost completely human in structure and to a large extent in mind before the period at which the races diverged from each other.
>
> (Darwin, 1998, p. 355)

This was not just a scientific hypothesis; it was a powerful human engagement. But it did have its own rather curious 19th century twist. Here is a remark on 'the essence of savagery':

> Nor is it an anomalous fact that the children of savages should exhibit a stronger tendency to protrude their lips when sulky than the children of civilised Europeans. For the essence of the savagery seems to consist in the retention of a primordial condition.
>
> (Darwin, 1998, p. 230)

I know some children of civilised Europeans that might have made Darwin think again.

Darwin also had a clear weakness for eugenic arguments:

> The weak members of civilised societies propagate their kind. No one who has attended to the breeding of domestic animals will doubt that this must be highly injurious to the race of man.
>
> (Darwin, 1871, part I, p. 168)

He also wrote that 'both sexes ought to refrain from marriage if in any marked degree inferior in body and mind' (Darwin, 1871, part II, p. 403), a sentiment that doubtless reflected his own anxiety about the possible risks he had run in marrying his own first cousin.

The experience of his *Beagle* voyage and the implacable determination of some of the European settlers to exterminate the indigenous inhabitants of Tierra del Fuego had left its mark too on Darwin's view of the likely future of inter-racial cooperation:

> At some future period not very distant as measured by centuries the civilised races of man will almost certain exterminate and replace throughout the world the savage races.
>
> (Darwin, 1871, part II, p. 201)

While it's clear that Darwin found this regrettable, even repellent, he also seemed to think there was little that could be done to prevent it. So the picture that emerges from Darwin's writings on race is certainly not one of a fully 21st-century sensibility clad in Victorian gentleman's garb. While he laid great emphasis on human cooperation he was at the same time easily seduced by eugenic arguments, albeit ones the full extent of whose consequences did not become clear until well after he died. There was also a charming element of parochialism:

> Hottentots and Negroes have readily become excellent musicians although they do not practice in their native countries anything that we should esteem as music.
>
> (Darwin, 1871, part II, p. 334)

I like to think that if Darwin was wrong, and there is an afterlife, he would now be getting lessons in music from the excellent Miriam Makeba who would have joined him recently.

I would like to close by saying something about the biology of gender differences today. Because of the sensitivity of the subject, it has been remarked much less than it should that human males and females are much more similar in body and brain size and behaviour than the theory of sexual selection would have led a neutral observer to expect. On Darwin's view of sexual selection nobody should expect the human races to be different in general talents and cognitive abilities. The whole point of sexual selection is that it works in the first place on relatively superficial characteristics such as skin colour and type of facial hair, while the sexual attractiveness of talent and ability should not vary much from one human population to another. So when modern researchers fail to find significant differences in talent between human populations, nobody should be very surprised.

But we ought to be very surprised how hard it is to find significant differences in overall cognitive abilities between men and women, given how common is sexual dimorphism in so many other species and given how adaptive it would have been for natural selection to save on expensive brain tissue in the human female if the main purpose of male brains was to impress women. That we can't find such differences is a really interesting fact, because it tells us that human males and human females were facing equally demanding cognitive challenges throughout all of our evolution. Indeed the subordinate and dependent condition of women that has characterised relatively recent centuries cannot have been true of the great bulk of our evolution since our common ancestor with the chimpanzees and bonobos. The history of writing on gender differences has been so politicised and so sensitive that we haven't drawn the appropriate conclusions from the fact that these differences are so hard to find.

I would suggest that if we look carefully we may be able to find more interesting gender differences in *preferences*: preferences for risk, for the kind of competitive environment that people like to inhabit, and for how they form networks. These are in no sense differences in general cognitive talents, but they may provide important clues as to why men and women still cluster very differently in their employment and their social life. For instance, in work I am conducting currently with Guido Friebel of the University of Frankfurt, we see that men and women behave differently in the way they use telephones (Friebel and Seabright, 2009). Women tend to make fewer calls than men but to make them last longer. This may be linked to differences in the way men and women form networks, with networks formed by women consisting of more strong ties but fewer weak ties. This remains a fascinating and so far underexplored area of research.

Conclusion

Charles Darwin saw, with clarity that escaped his contemporaries, that competition was everywhere in social life, but that it was not the antithesis of cooperation; the two were intimately intertwined. All primate societies are like this; they all embody cooperation within groups and competition between them. You struggle to be accepted by a group at the same time as you struggle to cooperate with other members of that group. As I suggested in that remark about the stamp of our lowly origin, I believe Darwin celebrated our primate nature. His own commitment to the unity of the human species was not a pious afterthought; it was a seamless part of the whole picture.

I'd like to leave you with that haunting picture from St Petersburg in Figure 4.2. As you look at it I'd like you to think that T. S. Eliot's vision of individuals lost in the mass, alienated and fearful, was missing something important about the human societies we know. These individuals are passing through; each of them leaves a tiny trace on the whole; but what they construct together is something greater than any individual could ever imagine. That is a legacy Charles Darwin would have invited us to appreciate.

5 The evolution of Utopia

STEVE JONES

Utopia is evolving. Since the place (or the non-place, for that is the real meaning of the word) was invented by Sir Thomas More in 1516, it has seen great changes. There have been many utopias since his fantastical work of prophecy and, for several centuries after that illustrious volume, the imaginary future did not alter much: in book after book, a new society, more religious, less greedy, or more rational, emerges (often from some cataclysm), although the people who live in it are little changed.

Towards the end of the nineteenth century, quite suddenly, the fictional future underwent a revolution. In 1895, H. G. Wells wrote *The Time Machine*. Its hero, the Time Traveller (who is never named) invents a device that projects him thousands of years forward. He makes a pause in AD 802 701, where he discovers a charming but ineffectual race of people, the Eloi, who − in the metaphorical Cambridge in which they live − eat fruit, inhabit ancient and decaying buildings and chat idly of large issues while not doing much to support themselves. Like Cambridge itself, the place struck the Traveller as effete, boring but just about tolerable.

Slowly, though, an unpleasant truth begins to emerge. Beneath the city's bland surface is hidden a pitiless world of industry and toil, peopled by a vicious and aggressive race, the Morlocks, who emerge at night to terrorise, and even to kill, the poor Eloi. Society, it seemed, had not changed much from that of Wells' own day, with a brutalised industrial class labouring to support their intellectual superiors. In *The Time Machine*, in contrast to earlier utopias, man himself has been revolutionised: so much so that

Darwin, eds. William Brown and Andrew C. Fabian. Published by Cambridge University Press. © Darwin College 2010.

Homo sapiens has evolved into two new and distinct species. In a twist which shows H. G. Wells to be a novelist rather than a literary hack, the Morlocks are in the end revealed as the masters, the intellectual Eloi as no more than their farm animals, to be fed, slaughtered and eaten as necessary. Life 800 000 years from now may still be bloodthirsty and unfair but those who live it have been transformed.

Wells' novel had a distinguished pedigree. Three decades earlier, evolution had entered the public imagination. Before *The Origin*, the only thing that seemed open to change was society. After it – and even more after its successor, *The Descent of Man*, published in 1871 – man himself suddenly became malleable. If *H. sapiens* had emerged from ancestors quite distinct from ourselves then surely, given time, he could change into something different again. With luck or with judgement, he might even be able to plan his own biological destiny. As Wells saw, why bother to imagine a new kind of society when it was more fun to invent one filled with new kinds of life? That evolutionary view is echoed in much of today's science fiction. The plots are identical to the horse operas and penny dreadfuls that once filled the shelves but their heroes and villains, as they live their tedious lives, take bizarre and eccentric physical form.

The inventor of the Morlocks was a supporter of Francis Galton, the founder of eugenics. If, members of that movement felt, bad genes were allowed to flourish at the expense of good, then perhaps our heirs would become a race of misfits. Wells' views were stark: at a meeting addressed by Sir Francis in 1905, six years before his death, he said that 'The way of Nature has always been to slay the hindmost, and there is still no other way, unless we can prevent the hindmost being born. It is in the sterilization of failures, and not the selection of successes for breeding, that the possibility of an improvement of the human stock lies.'

That idea reflected a shared concern of his day – and even of a century later – that Nature has become unnatural. She no longer slays the hindmost and, as a result, the race faces decay. The fear of evolution to come lives on. Few call for sterilisation today but there remains a vague disquiet that the best of our biology is behind us.

In my own view the opposite is true: that evolution is – at least in economically advanced countries and at least for the immediate future – well and truly over. Evolution, that is, as seen by H. G. Wells and his public; a change in physical attributes leading to the emergence of new and distinct forms of life, rather than in the more precise and banal scientific sense which sees it as no more than a change in gene frequencies. That latter process certainly continues. Even so, the notion that our very being is in a state of inevitable decline (or, for that matter, improvement) is wrong.

The Descent of Man says rather little about the descent of man (although it makes a powerful case that man did evolve and an even better one for the importance of sexual selection). In Darwin's day, so little was known about our own biology that the best that could be hoped for was to infer from other creatures what might drive human evolution. Technology has changed everything. *Homo sapiens* is now the creature of choice for biologists keen on Darwinism and we know more about our own biological history than about that of any other creature. Charles Lyell, Darwin's geological mentor, is famous for his phrase that 'the past is the key to the present'. The new insight into the human past also opens the door to our future.

Whenever it might happen, there is nothing mysterious about evolution. It is no more than genetics plus time. The process is inevitable. It depends on errors in reproduction. Any copy, be it of a painting or a gene, must be less than exact. A duplicate of a copy is, in turn, less perfect than what went before. To reproduce an original again and again, one replica succeeding the last, is to make – to evolve – something new. Darwin's great insight was to add a filter called natural selection to what he called 'descent with modification'. His idea is simple. It depends on inherited differences in the chance of having children. If one version of a particular gene is better at copying itself than are others and passes that ability on, it will replace what went before. Given enough genes and enough years, new forms of life will emerge.

Darwin's machine turns on variation, on shifts in the structure of DNA, on differences in the chances of survival and reproduction in a difficult and competitive environment, and on random variability in the chance of passing on genes in small populations. If we understand what is happening to those three activities today, we can guess what is in store for the future. In much of the world, the figures show, differences have been replaced by similarities – by a grand biological averaging that will drive our fate.

The legacy of the bombs

Human genetics has uncovered vast quantities of hidden diversity at the molecular level. Craig Venter's comparison of his own paternal and maternal genome copies with that of the reference sequence first used to read the double helix from end to end is an early hint at its extent. All such variation comes from mutation, most of which took place in the distant past. Venter found more than four million DNA variants even in that tiny sample. Most involved single-letter changes in the molecule, but almost a quarter were caused by insertions, deletions, multiplication of single short pieces or reversals of the order of blocks of genetic material. Few had a noticeable effect on the body (although Craig Venter bears genes that lead both to sticky ear wax and to a tendency to take risks). The figures suggest that one part of the DNA in 200 is likely to differ in the two chromosomal copies borne by every individual, a level five times higher than previously estimated. The 'thousand genomes' project, now under way, has set out to do the same across the world, picking up any variant present at a frequency greater than about one in a hundred. So rapid is technical progress that the job may be finished within little more than a year. Already it has revealed staggering amounts of diversity: so much so that every sperm and every egg made by every man and woman in the world is almost certainly different from every other. How much is relevant to adaptive evolution is hard to say but it puts to rest the once popular notion that every change in the DNA caused by mutation is likely to damage the purity of a homogeneous genome.

Even so, any increase in its incidence may be relevant to the future. Since the discovery of X-ray mutagenesis in the 1930s, and of the effects of chemicals a decade later, science fiction has seized on the idea that a poisoned environment will lead to a surge of mutated monsters (giant rats often play a part). What will happen to the mutation rate in years to come?

At the end of World War II, a team of geneticists arrived in Hiroshima. They expected – or feared – that the radiation from the bomb would damage both the genes of those who experienced it and, more alarming, those of their children. Certainly, an acute dose of high-energy rays killed thousands from radiation sickness as their DNA was destroyed, and in the longer term there was a slight increase in thyroid cancer in women exposed as children among the survivors. However, a massive survey of protein and DNA variation in the offspring of parents exposed to the bomb, and to a control group whose fathers and mothers had been further from the burst, revealed no difference in rate, which was a mildly reassuring surprise.

Another unexpected result soon emerged. Nine-tenths of the new mutations had taken place in the fathers, rather than the mothers, of the children being tested. This was a hint of what has since been established in other ways; that more genetic errors take place in the male than in the female line. Men and women make their sex cells in different ways. Women make all their eggs, in effect, before they are born, and release them at intervals throughout their reproductive lives. Men never rest, for they make sperm all the time. As a result, there are only about 20 cell divisions between the egg that makes a woman and the egg that she passes on, while for a father in his late twenties (the typical age of reproduction in the western world) there are some 300 divisions between the sperm that made him and the sperm he donates to the next generation. Each time a cell divides there is a chance of error, which means that many more mutations take place among males. Some argue that the effect is so great that most evolution among the great apes (the group to which we belong) has been driven by fathers. Whatever the truth of that claim, the most dangerous mutagens are men: and evolution is stuck with them for the foreseeable future.

Something about the first sex, though, is changing. Old men are even worse for the DNA than are young. Perhaps the most famous error in human genetics took place in the august testicles of Edward Duke of Kent. It gave rise to the haemophilia mutation passed on through his daughter Queen Victoria. The Duke was 51 when she was conceived; and in his case there were over 1000 cell divisions between sperm and sperm. For even older fathers, the figures are worse as the curve of damage rises steeply with age and (although its genetics are not straightforward) the incidence of autism is almost ten times higher in 50-year-old fathers than in those in their twenties. There are even claims that the IQ of the children of older fathers is lower than average, perhaps because of the accumulation of genetic damage.

Any change in the number of elderly fathers will hence alter the rate of mutation. Nowadays sex starts late but stops early. The average age of fathers in Britain is around 32 (which is about the same as in much of nineteenth-century Europe and colonial Canada). Crucially, though, there are fewer very old fathers today, for most people squeeze their reproductive lives, and their small families, into just five years or so, rather than the 20 or 30 once common. In Darwin's time, the mean age for a man to have his last child was almost 50 (and the great naturalist himself had his final offspring, who died young, at the age of 47). In modern Britain, only around 2% of all babies are born to fathers of 50-plus. The change in pattern is also manifest when comparing the third world with Europe: in French-speaking Cameroon, for example, fathers of 50 are, relative to others, the most fertile males, while in France itself 30-year-old fathers gain the gold medal.

Although most molecular mutations have little effect on the fitness of those who bear them, the decrease in the incidence of elderly male parents means that, if anything, the rate of error has probably gone down in the modern world compared to that in earlier centuries; and any concern of some genetic meltdown caused by the byproducts of modern times is misplaced.

Darwin's factory runs out of steam

Darwin's great insight was to see that life is a machine that makes copies of itself. Any mutation that improves its ability to do that job will spread as the generations succeed each other and may, in time, give rise to new kinds of creature. Natural selection, as he called it, is a factory for making almost impossible things. That approach – building upon inherited differences in the chances of reproduction – is now much used in industry. Since the 1960s, a new science of 'evolutionary computing' has emerged, based on strict Darwinian principles. It depends on designing a reasonably effective program and then randomly altering it. As various versions emerge, they are competed against each other for generation after generation. Only those that pass the efficiency test are allowed to reproduce. Over the generations, a much improved version of the initial program emerges, often working in a completely unpredicted manner.

Much as sceptics whine about intelligent design, natural selection made the eye, the ear and the elbow in exactly the same way.

It has also worked on the human race. One of the more remarkable contrasts of ourselves with our fellow apes is the striking differences in physical appearance across the world, most of all in skin colour. Darwin speculated that sexual selection – the preference for a mate with the same pigmentation as oneself – was involved; but he was wrong. The explanation is simpler and more direct. It turns on vitamin balance.

Almost everything we know about melanin – the dark substance found in black skin – is good. It protects against skin cancer (and Australia has ten times the rate as does Europe), slows the rate of ageing, reduces the damage to blood proteins and circulating vitamins (such as folate, crucial to pregnant women) caused by ultraviolet, and even improves the sight and hearing of those with dark skin and eyes. In one respect, though, it is lacking.

A shortage of vitamin D leads to the soft-bone disease called rickets. In the poorer parts of Darwin's London many children suffered from the condition and plenty died of it. Even today, rickets is the second commonest non-communicable disease of childhood. A deficiency of the vitamin also leads to an increased risk of multiple sclerosis, heart problems, certain cancers and infectious diseases. For most people, most of the substance is made through the action of ultraviolet on certain body chemicals rather than being taken from the diet. To do that the sunlight must first get in and melanin blocks it. A very fair-skinned person makes more than ten times as much as a typical African when exposed to the sun. Even in Britain, many people of African or Asian origin have low levels of the vitamin (a problem made worse by the black clothing and sheltered lives of many Asian women). As humans moved from the heat of the tropics to the grey skies of the north, any mutation that reduced the amount of melanin in the skin was favoured, and spread.

Several of the genes involved have now been found (the first, in homage to the power of the new genetics, as a melanin-free mutation in a certain fish). At least half a dozen genes, plus several more associated with variation in hair and eye colour, are involved. Remarkably, East Asian and European populations have lost their skin pigment in different ways, for separate genes in the biochemical pathway that makes melanin have been damaged by mutation in each of them. In each place, natural selection has had to wait for a new variant to turn up. It seized it at once, and with such energy that each skin-lightening gene spread rapidly through the population and dragged along a great segment of DNA on either side. Light skins must have evolved at least twice since humans left Africa – proof both of the power of selection and of its pragmatism: its ability to take up and utilise whatever random errors mutation might provide.

As well as the global trends in colour, north-west Europe contains a local patch of extraordinary pallor. Blonde or red hair, blue eyes and very pale skin were, before the mass migrations of historic times, found only there: and in northern Scandinavia such people formed as much as half the population. All this happened in just a few thousand years, with the spread of farming. That began in the Middle East some ten millennia ago, and moved at a surprisingly steady kilometre or so a year across the continent. Agriculture took 3000 years – around 100 human generations – to get to north-west Europe, and did not reach the rocky soils of northern Scandinavia until about the time of Christ. The first farmers moved from their diverse diet of wild flesh, fruits and grains (and their skeletons showed that they paid a heavy price for the shift, with plenty of evidence of deficiency diseases of various kinds).

Wheat, in its primitive form, can only be grown in places with a spring warm enough to allow the seeds to germinate. In most places the northern limit of cultivation is reached at a line that passes, roughly speaking, through modern Cambridge. The north and west of Europe, though, is warmed by the Gulf Stream, which means that grains will grow even in its grey and cloudy climate. Grains contain no vitamin D – and, worse, even adsorb the substance, should it be eaten in other foods. The northern farmers faced a vitamin deficiency crisis – and any mutation that further reduced skin pigment and allowed the stuff to be made in the limited sunlight available was favoured, even if it came at a cost (and Ireland, a nation of redheads, has one of the highest rates of skin cancer in the world). That pressing evolutionary problem – the origin of blondes – has been solved: and in spite of Darwin's conviction (shared by many people today) that it all has to do with sex, simple natural selection over just a few thousand years has done the job.

The process is hard at work in many other ways: in terms of diet, with only that minority of populations that took up cattle-herding able to digest milk in adulthood; in disease resistance to conditions as distinct as AIDS, tuberculosis and malaria; and even in shape and size. Some inherited illnesses of today (such as the lung disease cystic fibrosis) may be a relic of a time when those who bore single copies of the gene involved were better able to avoid infection. Medicine and social change have done a lot to reduce its power, with vitamin-supplemented food and drugs against most infections, not to speak of the joys of central heating and sun beds. Childcare has been important, too. In the old days – before World War II – the chance of a baby surviving was closely related to its weight. Underweight children died, but heavy babies died too. About half of all infant death was among children who were a kilogram below or a kilogram above the ideal weight. Nowadays, a kilogram too heavy or too light makes almost no difference to the chances of survival. As variation in birthweight is under the influence of genes, one more agent of evolutionary change has disappeared.

Once, only those who could run from sabre-toothed tigers or survive starvation stayed alive. Then it became a matter of fighting off cholera or the Black Death. Only those with genes to withstand them passed on their DNA. In much of the world those plagues have gone, as have tigers and starvation. Of course, plenty of people still die for genetic reasons – with strong inherited predispositions towards cancer, heart disease and the rest – but their fate is irrelevant to evolution, for such genes usually do not kill people until they are old enough to have passed them on.

To get a real insight into the course of natural selection we need information not from doctors, but from demographers (just like Darwin, for he came to the idea after reading Malthus' book on the growth of population). Because natural selection works only on differences, if everyone survived to the same age – however old or young – and had the same number of children – however many or few – then it would lose all its power. In the modern world we are rapidly moving towards a situation close to demographic equality. Across much of the globe, at least outside Africa, people are remarkably alike in their fate.

Ten thousand years ago, the struggle for existence really meant something. Skeletons from cave cemeteries show that most people died before they were 20 (although a few lived for three score years and ten). Even in Shakespeare's time, just one British baby in three survived to be 21 (and in Darwin's around one in two). Each death was raw material for evolution as it killed someone who might otherwise have passed on their genes.

Nowadays life is less tragic. Ninety-nine per cent of newborn babies in the affluent west live until puberty. For the first time in history, most people die old; perhaps as old as is possible. If, by some miracle, all infectious disease and all accidents were abolished by law, life expectancy would only go up by a couple of years. In the modern world, most of those who die have already had their chance to pass on their DNA. Their fate is of no interest to evolution.

Darwin's biological examination has two papers: the first, nowadays, easier than the second. To pass on genes one must stay alive (and presumably everyone reading these pages has done that); but then comes the difficult bit, finding a mate and having children. As Darwin showed in *The Descent of Man and Selection in Relation to Sex*, being male is a dangerous hobby. Males of most species compete with each other for access to females, and that can be dangerous. In some species the effect is extreme, with only about one male sea elephant in 20 surviving to adulthood, and a single successful male controlling up to a hundred females. For men, too, life is risky (and – a little known observation – they are several times more likely to be killed by lightning than are women). At birth there are about 105 males to 100 females, but this drops to near equality at the age of majority, and among 80-year-olds there are twice as many women as men. It probably has something to do with sex, for eunuchs and monks live for longer than do those con-demned to a normal sex life, while body builders and other abusers of testosterone tend to die violently, and young. That great universal is itself changing, for boy babies now survive almost as well as girls do. This means that there will soon be an excess of young men in those crucial years when they are searching for a mate.

Some of those young men will – as young men always have – commit evolutionary suicide in another way. Men are, of course, potentially able to have far more offspring than are women (if, that is, they can find enough mates); and, in the old days, some did – which meant that many more did not. Differences in male mating success are a potentially powerful agent of evolutionary change.

Nowadays, not many people are as fertile as they could be if they tried. Except in a few religious communities there are no women with 15 children. Men – most men, at least – are even more restrained, usually against their will. However, in ancient times, rulers were far more successful in their sexual lives than the ruled. Moulay Ismail the Cruel of Morocco had more than 800 offspring and perhaps a hundred mates (which meant that many of his male subjects had none). Such wild promiscuity leaves a mark in the genes, for the Y chromosome of a successful male can persist for many generations. That of the famously sexual Genghis Khan (and his equally promiscuous sons and grandsons) is said to be borne by sixteen million Asian men across what was once his empire.

Such behaviour might seem alien – but the imprint of ancient masculine philandering also marks the people of the British Isles. The southernmost half of Ireland, and the British mainland, each contain a diversity of Y chromosomes, with no male signifier unduly abundant at the expense of others. In the north-west corner of the island, though, in Donegal and nearby districts, as many as one man in five shares the same (or very similar) version of the Y. What is more, most of them carry one of a small group of surnames (inherited, of course, down the male line) such as Gallagher, Boyle and O'Donnell. Those families claim descent from the House of Niall, who were the High Kings of Ireland from the seventh to the eleventh centuries and each of whom descended from a perhaps mythical warlord, Niall of the Nine Hostages, Ireland's Genghis Khan (a man renowned for having kidnapped, among others, St David). Ireland had a male-dominated society that lasted until the sixteenth century, far later than in Great Britain. One patriarch, Lord Turlough O'Donnell, who died in 1423, had ten wives, 18 sons and 59 grandsons. His Y chromosome lives on, a witness to the potential power of sexual selection in *Homo sapiens*.

Nowadays sex, like death, has become average. Across Europe, the mean number of births per woman has gone down greatly in the past few decades, but – the important figure for natural selection – the variation in that figure has also decreased. Even in 1920, Russia had an average of around seven births per woman, while much of western Europe was at or below replacement level. On the global scale, most developed and near-developed countries, from Germany even to China, have rather similar rates, with no more than double the overall rate in China compared to Germany and rather little variation among families within each country. Outside our natal continent, many countries have reached, or are moving towards, a new way of life, with low rates of death matched by similarly low rates of birth. Once again, Africa is the great exception, for countries such as Niger and Uganda have five times the typical European rate of birth (and parts of the Middle East are also having a baby boom).

The figures for variation in survival and in reproductive success can be combined to give a measure called 'the opportunity for natural selection'. India is a microcosm of what is happening across most of the world. It contains affluent members of the middle class, a vast peasant population not very different in its way of life from that of eighteenth-century Europe and a few remaining tribal peoples in the foothills of the Himalayas. Putting all the data together shows that a Grand Mediocrity – everyone more or less the same in survival and in sexual performance – is looming for us all. It has already caused natural selection to lose 80% of its power in bourgeois India compared to what it is in the tribes.

As a result, Darwin's grand factory for improbability is running low on fuel: and as long as the modern world remains a relatively comfortable and sexually calm place, its capacity will remain a shadow of what it once was. In spite of the undoubted fact that modern medicine is allowing some genetically feeble people to survive when once they would have died young, natural selection, if it has not surrendered to human ingenuity, has at least declared an armed truce.

Out of Africa and into the future

We are all Africans but some of us are more African than others. Humans escaped from the continent of their birth some 65 000 years ago into the Middle East. From there they spread, in the end reaching the Americas around 20 000 years before the present day and New Zealand only about 1000 years ago. The genes show a great split between Africa and the rest of the world (and a lesser split within Africa itself, where two groups of hunters, the Pygmies and the speakers of Khoesan languages in the south, are on a distinct and ancient branch of the family tree), for the teeming peoples of Asia, Europe and the Americas bear but a small sample of the genes of our natal continent, proof that only a few made it across the bridge to the outside world. By comparing the extent of variation in the source population with that in its descendants it is possible to make an educated guess about the size of the migrating group (and – a moment of self-aggrandisement – the first estimate, made in 1986 by myself and a physicist colleague, based on the patterning of globin genes within and outside Africa, was of six individuals over 200 years or of rather more if it lasted for longer). Comparison of patterns of variation on the X chromosome (most of which are borne by females) and the autosomes (shared equally by the two sexes) now suggest a smaller bottleneck for the former, as a strong hint that more men than women took the risk of escaping from their continent of origin. Later, though, as humans filled the world, about equal numbers of each sex seem to have made the trip to new lands.

Each time a population goes through such a bottleneck, the accidents of sampling mean that some genes are lost. As a result the daughter population is less diverse than its parent. A map of the extent of human genetic variation in relation to distance from Addis Ababa (close to where the presumed centre of dispersal was) shows a precise fit, with the least variable populations at the tips of the branches of humankind's family tree, in the southern part of South America, in New Zealand, and on remote Pacific islands. A history of accidental change in small populations has driven much of our biological past.

It is – or was until recently – hard at work on a smaller scale. Humans are different from other primates in one obvious way; they are enormously abundant. Compared to other mammals of about our body size we are 10000 times more common than we would be if we followed the same ecological rules as do deer or porpoises. Such population growth began with farming and within the past century has become an explosion of abundance – so much so that most of those reading these pages will have seen more people on the way to work than the average hunter–gatherer met in a lifetime. Our relatively low levels of DNA variation in comparison to chimpanzees and gorillas also hints that, for much of the past, *Homo sapiens* was an endangered species.

Rare creatures live in a permanent bottleneck. They are forced to mate with those who share a recent ancestor, for want of anyone better. Such inbreeding leads to a loss of genetic diversity and to the divergence of isolated populations as a result of evolution by accident. Populations marooned on islands, real or metaphorical, tend to go in for it. European royalty are enthusiasts; but, to a degree, so are we all. Like it or not, each of us marries someone from our own family, because we have no choice. It is a matter of arithmetic: if everyone had two parents, four grandparents, eight great-grandparents and so on, quite soon the numbers become huge and we run out of possible ancestors. The answer is simple: we share them with our spouses.

The common ancestor of any pair of kissing cousins lived just two generations ago. In fact, every Briton of European descent is, roughly speaking, a sixth cousin to all the others; their shared predecessor was around when Darwin was a young man. But how far back must we go to find the common ancestor of everyone; the person that all Africans, Europeans and Tibetans can claim as a distant grandparent? He or she may have lived just 2300 years ago – the time of Alexander the Great. To get to the universal ancestors (when everyone alive was the forefather or mother of everybody alive today, or of nobody), we need travel back only 5000 years – the era of the first pyramids. Had you then entered any village on Earth, the first person you met would, if he or she had living heirs, trace descent straight to you, and to everybody else.

Everyone is, as a result, to some degree inbred. Does that matter? For incest, certainly: the children of such matings bear a heavy burden of inborn disease. For cousins the figures are less clear, but in Europe there is a small but noticeable increase in the death rate and a matching decline in health of infants, while in the Middle East, where certain blood diseases are common and cousin marriage has long been customary, the figures are worse. The evolutionary effect is also important, for – as the African bottleneck shows – isolated and inbred populations can diverge from each other by accident.

Cousins are at risk of having an unhealthy child because, should their shared grandparent bear a gene harmful in double dose, each has a chance of receiving a copy that might then be passed on. The same is true of all their tens of millions of inherited variants, harmless or not, scattered along the DNA, often in great blocks. As a result, a scan of how much of anyone's DNA is present in double dose, identical on the two versions of each chromosome, one from the father and one from the mother, hints how inbred they might be.

Everyone has plenty of short paired bits of double helix, perhaps from cousin marriage long ago, but in some places the effect is extreme. The northern part of Orkney tells the tale. The islands have, now and again, gone through population crashes as the weather turned bad or the economy collapsed (and numbers have declined by two-thirds in the past century and a half). People were, as a result, often obliged to marry a relative through lack of choice. A search for paired 10 000-letter long segments of identical DNA along a chromosome gives strong evidence of shared ancestry. One in four native Orcadians had at least one such lengthy and doubled-up stretch. A comparison of their incidence with marriage records shows an almost exact fit. The result may impress those in search of an exclusive and perhaps aristocratic family tree; but not all the news is good. Patients with diseases as different as schizophrenia and colon cancer are more likely to carry such long genetic signatures, perhaps as evidence that ancient within-family sex affects a descendant's health.

Such kin mating is on the way out – which is good for world health but will much reduce the chances of evolution by accident. The effect is obvious when Orkney is compared with the mainland of Scotland, for the proportion of those with double copies of long stretches of invariant DNA is only one-fiftieth of that on the islands, proof that they have married outside their family. In the United States, for example, elderly people of European origin – those born near the beginning of the last century – have, when compared to teenagers of the same background, many more such doubled lengths; proof that even in the lifetime of the present generation, a decline in inbreeding has led to rapid population mixing. Within no more than a couple of centuries at that rate, the nation (or at least its European inhabitants) will in effect form a single large outbred population. With the development of great cities and the onset of mass movement the effect is universal.

All over the world, people, and genes, are on the move. The spread of surnames – with the names (and the DNA) of the High Kings of Ireland now more abundant in North America than in Donegal – shows the strength of the effect. My own workplace, University College London, has students from over 150 countries. I would not dream of prying into their private lives, but Nature has a way of taking its course and no doubt many will settle down together. So strong is the healing power of lust that, in Britain as a whole, and excluding religious minorities, education level now predicts who a young person will marry better than does skin colour. Once again, a major part of the Darwinian machine is less able to generate divergence than once it could.

The evolution of Utopia

Darwin had a wonderful phrase to describe those who were ignorant of his theory: they 'look at an organic being as a savage looks at a ship, as at something wholly beyond his comprehension'. No doubt our descendants may some day look back upon a twenty-first-century way of life that for them is beyond comprehension in its ugliness: but however amazed our successors may be at how far from a social Utopia we were in those benighted times, and however wonderful the advances that medicine or education will bring, they will look pretty much like us. That is because for human beings, uniquely, the evolutionary future lies not in what we are, but in what we think.

In fact that has long been true. The strangest thing about human evolution is how little there has been. No other species is so widespread and none fills so many gaps in the economy of Nature. Many animals carry out tasks as wonderful as some of our own but through biology rather than intellect. For them, success at one job means failure at all others. In the past hundred thousand – in the past hundred – years, our lives have been transformed but our bodies have not. We did not evolve much, because society did it for us. As the world becomes more socially homogeneous that grand averaging will continue and – although we will change as populations mix and mate, the notion of some inevitable decline (or even progress) is even less likely than in H. G. Wells' time.

Our more distant prospects are less hopeful. In *The Time Machine*, the Traveller journeys on, further and further. At thirty million years he has a vision of a land peopled by red, crab-like creatures and then, at last, of a moribund Earth as the Sun grows cold and all life comes to an end. Although its author was wrong about the details of how and why it will happen he was dead right about our inevitable extinction. Darwin himself, in a letter of 1865, also saw a long-term apocalypse (and commented on the leaden pace of human evolution):

> I quite agree how humiliating the slow progress of man is; but everyone has his own pet horror, & this slow progress, or even personal annihilation sinks in my mind into insignificance compared with the idea, or rather I presume certainty, of the sun some day cooling & we all freezing. To think of the progress of millions of years, with every continent swarming with good & enlightened men all ending in this; & with probably no fresh start until this our own planetary system has been again converted into red-hot gas.— *Sic transit gloria mundi*, with a vengeance.

In Dante's *Inferno* a special punishment is reserved for those who dare to divine the future: their heads are twisted around on their necks so that they are forced to look backwards and to study only the past. Even so, in this most auspicious Darwinian year, I will stick my own neck out and suggest that – in the developed world and for the time being at least – those who worry that the glory of the world will soon pass should cheer up: because, as far as the biological Utopia is concerned, they are living in it now.

6 The making of the fittest: the DNA record of evolution

SEAN B. CARROLL

Introduction: A second golden age

Just 200 years ago, when Charles Darwin was born, most of the world was an unexplored wilderness. The creatures, plants and people that inhabited the lands beyond Europe were largely unknown. So, too, was the history of the planet and of our own species.

By the time Darwin died in 1882, the explorations of many previously unseen parts of the world and the unearthing of the history of life had sparked a revolution that profoundly changed our perception of the living world and humans' place in it.

Three great voyages were pivotal to this revolution in thought. The best known, of course, is Darwin's five-year journey around the world on the *HMS Beagle* (1831–6). But the voyages taken by Alfred Russel Wallace, first up the Amazon River (1848–52) and then through the Malay Archipelago (1854–62), and of Henry Walter Bates (who spent 11 years in the Amazon (1848–59), 10 of these on his own) made seminal contributions to the birth and growth of the theory of evolution by natural selection.

Darwin's, Wallace's and Bates' voyages and works mark a truly golden age of exploration and discovery. Many of their insights stemmed from their prodigious efforts at collecting the plants, animals and fossils of the regions they bravely explored. Through the study of certain creatures, such as the finches and tortoises of the Galapagos Islands, and the butterflies of the Amazon and the islands of the Malay Archipelago, Darwin and Wallace established the fact that species changed and

Darwin, eds. William Brown and Andrew C. Fabian. Published by Cambridge University Press. © Darwin College 2010.

explained that life was a struggle for existence, with only the most fit surviving and reproducing.

But none of these great naturalists knew *how the fittest are made* – how evolution works at the most fundamental level to generate the adaptations that enable creatures to thrive in various habitats and that make species different from one another.

Now, some 150 years after *On the Origin of Species* (Darwin, 1859), we can do just that. I believe that this newly acquired power has spawned a second golden age of evolutionary science. Modern biologists are collectors, as well, but instead of pickling and stuffing creatures for museums, we are collecting DNA from them and assembling a massive new DNA record of evolution.

In just the past decade or so, we have obtained the entire DNA sequences from more than 2000 species, including more than 50 mammals, and partial DNA sequences from thousands more creatures. Inside this DNA record is vivid documentation of, and some surprising revelations about, the evolutionary process. A whole new set of creatures is emerging as icons of evolution. One of these new icons, and perhaps the most remarkable animal I know of, first came to light on a voyage many decades ago to one of the far-off corners of the globe.

The icefish

In the 1920s, commercial whaling was thriving. The Norwegian fleet and government wanted to find new stocks of whales so they sent research ships to the Southern Ocean. On 14 September 1927, the converted wooden sealing boat the *Norvegia* went to sea with three primary missions (Norske videnskaps-akademi i Oslo, 1935).

The first objective was to voyage to remote Bouvet Island, a tiny speck in the Southern Ocean, in order to establish an outpost there that could assist the whaling fleet. The second mission was to claim the island for the government of Norway. And the third assignment was to conduct research on marine life in order to better understand the whale population and the food chain in these waters.

After a 6000 mile voyage and about three months at sea, the *Norvegia* reached Bouvet Island and the landing party erected a pole bearing the

FIGURE 6.1 Adult mackerel icefish *(Champsocephalus gunnari).*

Norwegian flag. The first two missions accomplished, the crew then spent the next month exploring the waters around Bouvet.

It was a tradition in the Norwegian fleet to have zoology students aboard both research and commercial ships. The young zoologist aboard the *Norvegia*, Ditlef Rustad, spent as much time as he could trawling for creatures. On the day after Christmas, at a depth of about 100 feet, he caught a very unusual looking fish (Figure 6.1). It had big eyes and large pectoral and tail fins like other fish, but it also had a long protruding jaw full of teeth, and was very pale, almost transparent. He dubbed it a 'white crocodile fish'.

Rustad then did what any good fisherman would do – he filleted it – and much to his surprise he discovered that, unlike any other fish he had dissected, its blood was completely colourless. He took some pictures of his white crocodile fish. The *Norvegia* soon continued its journey to other destinations and eventually returned to Norway. The fish was long out of Rustad's mind, until two years later, when another zoology student mentioned a bloodless fish.

Johan Ruud had just returned from his own voyage to the Antarctic aboard a factory ship. In the course of his journey, a few of the crew told him to keep an eye out for fish with no blood. At the time, Ruud thought that the old hands were just joking with him and later shared the yarn with Rustad. But Ruud was stunned when Rustad told him, 'I have seen such a fish', and showed him the photographs (Ruud, 1965).

Ruud heard nothing more about the crocodile fish or 'icefish', as the whalers also called them, until 1948. One of his own students returned from an Antarctic expedition with some preserved icefish. They had white gills, unlike the red gills of other fish. Ruud's curiosity was re-ignited and he became determined to get to the bottom of the mystery of this supposedly bloodless fish.

In 1953, nearly 25 years after he first heard tales of the creature, Ruud returned to the Southern Ocean and set up a lab on South Georgia Island. A few icefish were brought to him and as he drew their blood, he saw right away that it was almost transparent. Later, under the microscope, he confirmed what he thought was impossible – the blood contained no red blood cells. The icefish, unlike any other fish, completely lacked pigmented oxygen-carrying cells.

Ruud reported his discovery in *Nature* in 1954 (Ruud, 1954). He noted in his paper, 'The fact that a good-sized vertebrate can exist without any oxygen-binding blood pigment raises some interesting questions'. The first question was 'What on earth would cause them to abandon a way of life that had nurtured their ancestors for 500 million years?'

Ruud surmised that the dramatic difference in icefish physiology must be linked to the exceptionally cold habitat in which icefish live. The waters around the Antarctic average less than $29\,°F$ ($-2\,°C$). Ruud noted that the icefish blood was extremely dilute, about 99% liquid by volume and only 1% cells (all white cells), whereas red-blooded fish relatives have 20% or more cellular blood volume. The dilute icefish blood would certainly be less viscous and easier to pump in the extreme cold.

How then did this fish evolve? What happened to their haemoglobin and how could the fish survive without it?

The typical place biologists would begin to explore the origin of a group of species would be the fossil record. However, there is no record of the icefish, and even if we had fossils, we would not be able to tell from mere bones what colour their blood was and when or how it changed. But there is now a record of the history of the icefish that is accessible – in their DNA.

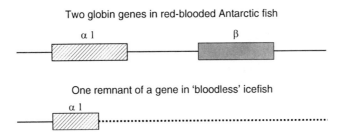

Two globin genes in red-blooded Antarctic fish

One remnant of a gene in 'bloodless' icefish

FIGURE 6.2 Globin genes are lost from or fossilised in icefish. Top, in red-blooded fish, the genes encoding the two different protein chains of the adult haemoglobin molecule are adjacent to one another. Bottom, in icefish DNA, the β-globin gene has been lost and the α-globin gene is a truncated molecular fossil.

The DNA record of the icefish reveals what was lost, modified and gained in the course of their evolution. I will focus first on what was lost – their haemoglobin.

The mystery of the absence of haemoglobin in icefish had to wait until more than 40 years after Ruud's first description of the bloodless fish. In 1995, Cocca *et al.* (1995) probed icefish DNA for the two genes that encode the protein chains of the haemoglobin molecule, their adult α- and β-globin genes. In red-blooded fish, the two genes are located adjacent to one another in a head-to-head orientation. But in all icefish species examined, the β-globin sequence has vanished – it is extinct (Figure 6.2). And the α-globin gene is just a truncated remnant, a molecular fossil – it still resides in the DNA of the icefish, but it is completely useless and eroding away just as a fossil withers upon exposure to the elements. The fossil gene is vivid proof that the icefish has abandoned a haemoglobin-dependent lifestyle that all other vertebrates depend upon.

Of course, evolution is not all about losses – it is also about invention. In the DNA record of the icefish there is also rich information about how these fish have adapted to the very cold waters of the Southern Ocean.

The low temperature of Antarctic waters ($<29\,°F$; $-2\,°C$) presents some considerable challenges to body physiology. In addition to large icebergs and icesheets, the water contains lots of ice particles. The danger these particles pose to fish is that, if ingested, they can nucleate freezing of stomach and body fluids and – bang – they're fishsticks! Most tropical and temperate fish freeze at about 30.5 °F (−1 °C), so Antarctic fish must have evolved some means to cope with the icy ocean.

One group of fish, the suborder *Notothenioidei*, of which icefish constitute just one family, dominates the Antarctic waters, making up one-third of all Antarctic fish species and most of the fish biomass. The first clues to how notothenioids have adapted to the cold came from a series of studies in the late 1960s which showed that their blood serum froze at lower temperatures than other fish and contained high levels of antifreeze glycoproteins (AFGPs) (Somero and DeVries, 1967; DeVries and Wohlschlag, 1969; DeVries, 1971). Subsequent structural studies revealed that AFGPs have a very unusual and simple structure consisting of a repeating tripeptide (theanine–alanine–alanine or theanine–proline–alanine) with a sugar attached to the theanine residue. The AFGPs work by absorbing small ice crystals and inhibiting their growth.

Since warm-water fish do not have AFGPs, the antifreeze is a notothenioid invention. Analysis of AFGP gene structure has revealed how the antifreeze evolved from part of another, functionally unrelated gene encoding a trypsinogen-like digestive enzyme. The three-amino acid repeat of the AFGP evolved from a nine-base pair sequence in the trypsinogen-like gene. The repeat was duplicated and then repeated many times to form the repeating structure of the gene and protein (Chen *et al.*, 1997; Cheng and Chen, 1999).

The very clear forensic DNA record of the origin of the antifreeze gene is an outstanding example of how evolution works most often by tinkering with already existing materials – in this case parts of an existing gene – rather than by designing new molecules from scratch.

Other icefish genes reflect more subtle modifications that have enabled these extraordinary creatures to adapt to the extreme Antarctic environment.

The DNA record of the icefish (and of all species) reveals three general features about the making of the fittest:

1. The Earth and life evolve together. Species are adapting to a changing planet. Invasions of new habitats require changes in lifestyle – in form and physiology.
2. Shifts in lifestyle are reflected in the DNA sequences of pertinent genes. In the course of adaptation, some genes are born, some genes change,

and some genes, like the globin genes of icefish, die when organisms shift their habitats.

3.   The DNA record is rich documentary evidence of Darwin's principles of natural selection.

Let's revisit those principles to examine just exactly how Darwin defined natural selection.

The 'great and complex battle of life' leaves a record in DNA

The most succinct explanation of Darwin's conception of the evolutionary process was offered in Chapter IV of *On the Origin of Species* (Darwin, 1859) (I have italicised key terms and phrases):

> Can it, then, be thought improbable . . . that other *variations* useful in some way to each being in the *great and complex battle of life*, should sometimes occur in the cause of *thousands of generations*? If such do occur, can we doubt (remembering that many more individuals are born than can possibly survive) that individuals having an *advantage*, however slight, over others, would have the best chance of surviving and of procreating their kind? On the other hand, we may feel sure that any variation in the least degree injurious would be rigidly destroyed. *This preservation of favourable variations and the rejection of injurious variations I call Natural Selection.*

The key elements to natural selection then are variation, a slight advantage, and the compounding effect of time. But Darwin did not know the basis of variation or how large or small an advantage was necessary for the process to unfold. He did realise that there were two faces of natural selection. While the preservation of favourable variations garners the most attention from biologists and is the concept most people grasp about evolution, the rejection of injurious variations is equally important. Furthermore, it will be seen in the examples I discuss below that whether a variation is injurious or advantageous (or neither) depends upon the conditions faced by a population of organisms – the fittest is very much a conditional status.

Over the past couple of decades, biologists have deciphered the genetic and molecular basis of many kinds of changes within and between species. For the remainder of this essay, I am going to focus on just one dimension of Darwin's 'great and complex battle of life' – the battle between seeing and not being seen.

In the animal kingdom, the contest between predator and prey has been ongoing for more than 500 million years. Some of the most obvious aspects of this battle are the devices that prey use to evade their hunters, and the countermeasures that predators have evolved to detect prey.

The battle between seeing and not being seen

One animal battlefield that has yielded an exquisite model of natural selection is the desert of the south-western United States. Over the past million years or so, there have been several episodes of volcanic activity that have produced long lava flows over the scrub desert. Different types of habitats, open scrub and lava rocks, impose different demands on the animals living within them. One resident of these habitats, the rock pocket mouse, is emerging as a new icon of the evolutionary process.

These desert mice occur in two colour forms, or morphs: a sandy-coloured form and a dark 'melanic' form (Figure 6.3). Michael Nachman of the University of Arizona and his colleagues have shown that sandy-coloured mice are found most often on sandy-coloured soil and rocks, and dark-coloured mice are found most often on black lava rocks. This is not because the animals know which type of habitat conceals them better from their various predators such as birds, snakes and lizards. Rather, the predators cull the mismatched animals. It is very clear from decades of observation that natural selection is operating on mouse fur colour.

Moreover, Nachman and colleagues have been able to determine the precise genetic basis of the colour difference. There is a perfect correlation between the presence of four mutations in the *melanocortin-1 receptor* (*MC1R*) gene of dark mice that results in four amino acid changes in the MC1R protein relative to the sequence of the protein in sandy-coloured mice (Nachman *et al.*, 2003). These mutations cause the fur to darken.

The scenario for the evolution of the dark rock pocket mice then is that a new habitat was presented by the lava flows. Think of the lava flows as being much like the volcanic Galapagos Islands, only in this case the islands are in a sea of sandy desert. The pocket mice are seed gatherers

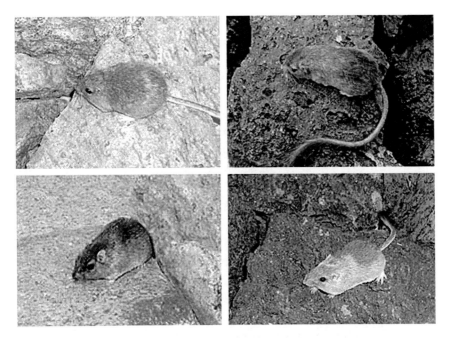

FIGURE 6.3 Association of habitat and fur colour in rock pocket mice. Light-coloured mice are typically found on light-coloured rocks and dark mice on dark lava flows, which affords them protection from predators. Photos courtesy of Michael Nachman, University of Arizona, from Nachman *et al.*, 2003.

and as the lava breaks down, the soil provides a nutritious substrate for plants. But in order to invade the new habitat and to exploit its resources, a change was necessary in the ancestral population of sandy-coloured mice – the darkening of fur to enable populations to blend with the dark habitat. Mutations in *MC1R* were then favoured in populations that invaded the lava flows. It is crucial to appreciate that the advantage of the *MC1R* mutations is entirely conditional. In open sandy-coloured habitat, dark mice are easy prey so such mutations would be injurious. Which fur colour is most fit depends on where a mouse lives.

In other settings, it is the predator that adapts to the challenge of the environment. Consider the European kestrel, whose favourite food is the vole. This bird must locate its small prey among the overgrown fields of northern Europe. How does a bird flying at some altitude pick out a small vole among the vegetation?

The kestrel has evolved a nifty adaptation involving its sense of colour vision. In humans, what we describe as full colour vision is our ability to detect light at wavelengths from about 400 nanometers (nm) (violet) to about 700 nm (red). We do so through three proteins in our retinas called opsins that are excited by different wavelengths of light. The colours of objects are due to the wavelengths of light that are absorbed or reflected, and this depends upon their molecular composition. Our three opsins, which are also present in great apes and old world monkeys, allow us to see a broader colour spectrum than other mammals, which generally have two or just one opsin.

Birds, however, do have full colour vision. One opsin in birds is tuned to shorter wavelengths around 400 nm, in the violet range. But in kestrels a mutation has happened in this opsin that shifts its absorption maximum by 35–38 nm, a very dramatic shift. The mutation causes the replacement of a serine residue with a cysteine residue. The resulting sensitivity of the bird opsin is near 370 nm, not in the violet range but in the *ultraviolet* (Shi and Yokoyama, 2003; Ödeen and Håstad, 2003). In most animals this mutation would be injurious.

But not in kestrels. Why? Because vole urine reflects in the ultraviolet (Viitala *et al.*, 1995). This means that whenever voles are active, on trails or around burrows, there are markings that the kestrel can perceive from altitude. The kestrel then need only concentrate its search where the ultraviolet markings of the voles are abundant.

It is crucial to emphasise that in the case of the *MC1R* and opsin genes, and in all natural genetic variation, the functional mutations arise at random. The selective conditions (habitat, predators, mates, etc.) determine which variants are favoured – which are the *fittest*.

Use it or lose it

One might think that once creatures have evolved nifty capabilities such as colour vision or ultraviolet vision, that they would never relinquish these abilities. But that is not the case. As species evolve to keep up with the changing Earth, or as they leave one habitat and invade another, they often adapt a lifestyle that is different from that of their ancestors. These habitat and lifestyle shifts leave a record in DNA.

Consider, for example, two fish species living in different zones in the ocean. A shallow-water reef-dwelling fish, such as a clown fish, lives in brightly lit waters where all of the wavelengths of light are able to penetrate. The bright colours of reef fish are used in communication, mimicry and camouflage, and these fish have full colour vision and a full set of opsins. But a deeper-water fish such as the coelacanth lives in a much darker habitat, reached only by dim blue light, at most.

The coelacanth's dim light lifestyle is reflected in its opsin genes. It does have one short wavelength opsin; however, the code of this coelacanth gene is altered in several places that disrupt its function (Yokoyama et al., 1999; Shi and Yokoyama, 2003). The coelacanth opsin gene is a fossil. Some ancestor of the coelacanth must have had a functional short wavelength opsin but it was inactivated at some time in the course of coelacanth evolution.

Why would a good functional gene be allowed to accumulate mutations and to become a fossil? Because that gene is of no consequence to coelacanth physiology at the depth where it lives. Mutations in the opsin are neither advantageous nor injurious. Therefore, natural selection is 'blind' to the state of the coelacanth opsin. When selection is released on a gene, mutations can accumulate in its DNA code without any effect on the organism. One apt description of this principle is 'use it or lose it'.

This is clearly the same situation for the erosion and loss of globin genes in icefish. With no red blood cells, mutations in the globin genes are of no consequence. One may wonder then, are fossil genes just some rare quirk of odd animals such as icefish and coelacanths? Not at all. Fossil genes are exactly what we would predict to evolve as a consequence of the continuing action of mutation, over time, in the absence of natural selection.

We can see this principle in every species. For example, in the nocturnal owl monkey, its short wavelength opsin is also inactive (Jacobs et al., 1996). This gene is intact in all of the diurnal (daylight-active) relatives of the owl monkey, so it is quite reasonable to deduce that the shift to a nocturnal lifestyle relaxed selection on colour vision and this opsin gene. The same gene has been inactivated in other nocturnal animals such as the bush baby (Jacobs et al., 1996), and in the subterranean blind

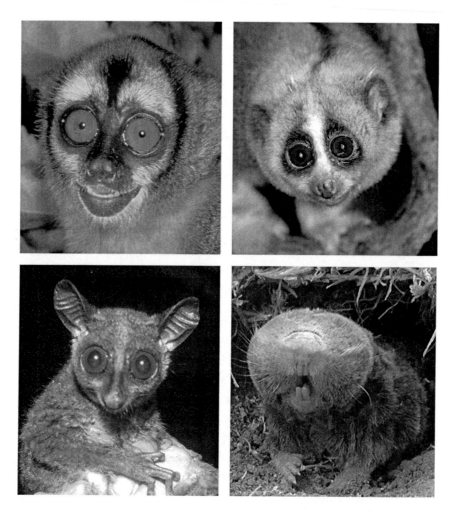

FIGURE 6.4 Nocturnal or subterranean mammals with fossilised opsin genes. The owl monkey (top left, photo by Greg and Mary Beth Dimijian), bush baby (bottom left, photo by B. Smith, Cercopan, Nigeria), slow loris (top right, photo by Larry P. Tackett, www.tacketproductions.com) and blind mole rat (bottom right, photo by Tali Kimchi) all have fossilised *SWS* opsin genes as a result of their adaptation to nocturnal and subterranean lifestyles. Montage from Carroll (2006).

mole rat (David-Gray *et al.*, 2002), a creature that has adopted an underground lifestyle (Figure 6.4).

It is important to stress that in the coelacanth, the owl monkey, the bush baby and the blind mole rat, the inactivating mutations are different. It is clear that the opsin gene has been inactivated independently many times in animal evolution as a consequence of shifting to darker habitats.

Fossil genes are prevalent in humans as well. Hundreds of genes that were active at some time in our vertebrate mammalian, and primate, ancestors no longer function in humans. In all species, fossil genes tell us about how current species' lifestyles are different from those of their ancestors. Fossil genes also demonstrate that evolution is not a progressive process – some genetic information gets lost over time that cannot be re-acquired. Modern species are not 'better' than their ancestors, they are just different.

Moreover, fossil genes are excellent evidence against notions of purposeful 'design' in the making of species. What designer would design creatures with hundreds of inactive genes? The patterns of gene gain and loss seen in species' DNA are exactly what we should expect to see if natural selection acts only in the present, and not as an engineer or designer would. Natural selection has no way to preserve genetic information that is not being used. It also cannot plan for the future. The inactivation and loss of globin genes or opsin genes is a one-way process – organisms cannot hold on to unused genes 'just in case' they might be needed some day.

Evolution is reproducible

There is another very important principle revealed by the independent inactivation of opsin genes in creatures that live in dim or dark habitats – similar events can and do repeat themselves in DNA.

This is true not only of the fossilisation of genes but of functional changes in genes as well. For instance, the *MC1R* gene is not only responsible for melanism in rock pocket mice, but in a host of other animals, including the bananaquit, the lesser snow goose, jaguars and the Arctic skua (see Carroll, 2006 and references therein). In the latter case, the very same mutation occurred as in the rock pocket mouse.

Many other examples of the repetition of evolution in different species, either in the same gene or even in the very same base pair, have been documented. It is important to appreciate that, prior to the emergence of the DNA record, there was no way of knowing whether similar changes in different species had a common cause. *The repetition of evolution shows how similar selective conditions favour similar genetic variations in different species, at different times, in different parts of the world.*

A golden age

The new, massive DNA record of evolution endows modern biologists with powers that Darwin could not have imagined. We can now see how the diversity of life is made. We can pinpoint the exact genetic changes that enable organisms to adapt to different conditions, we can decipher how current species are different from their ancestors, and we can determine the relationships among species in the tree of life. We may now even *predict*, given a set of conditions, the genetic basis of evolutionary changes in particular traits.

The power of the DNA record is also being harnessed to address the major issue that Darwin sidestepped in *On the Origin of Species* – the origin of humans. The DNA records of modern humans, when compared with those of chimpanzees, gorillas and other apes, and remarkably, of our closely related *extinct* cousins, the Neanderthals, promise to reveal just how similar we are to other species, and how we evolved to be unique. Another great voyage lies ahead.

7 **Evolutionary biogeography and conservation on a rapidly changing planet: building on Darwin's vision**

CRAIG MORITZ AND ANA CAROLINA CARNAVAL

> As buds give rise by growth to fresh buds, and these, if vigorous, branch out and overtop on all sides many a feebler branch, so by generations I believe it has been with the Great Tree of Life, which fills with its dead and broken branches the crust of the earth, and covers the surface with its ever branching and beautiful ramifications (Darwin, 1859, p. 130).

Biogeography, the analysis of the geographic distribution of biological diversity, has roots extended to the Enlightenment and beyond (Lomolino *et al.*, 2004). Yet the discipline was revolutionised by, and highly influential to, the field of evolutionary biology set forth by Charles Darwin and his contemporaries – particularly Alfred Russel Wallace. Darwin and Wallace built on earlier studies (e.g. Hooker, 1844; Lyell, 1830), and on detailed observations of the natural history and geology of whole biomes, to establish the core principles of what we now refer to as 'evolutionary biogeography' (Darwin, 1859; Wallace, 1852; 1855; 1860; 1865; Figure 7.1). Importantly, each of these remarkable scientists combined extensive field observations with a deep knowledge of particular groups of organisms to generate new insights into how evolution – i.e. descent with modification – interacts with environmental variation to shape the geographic distribution of diversity. These concepts, with subsequent elaboration, remain central to our understanding of how biological diversity is generated (Figure 7.1).

Now, more than ever, we must integrate our knowledge of evolutionary processes and biological patterns to protect biodiversity in the face of an unprecedented global change imposed by a single species – our own.

Darwin, eds. William Brown and Andrew C. Fabian. Published by Cambridge University Press. © Darwin College 2010.

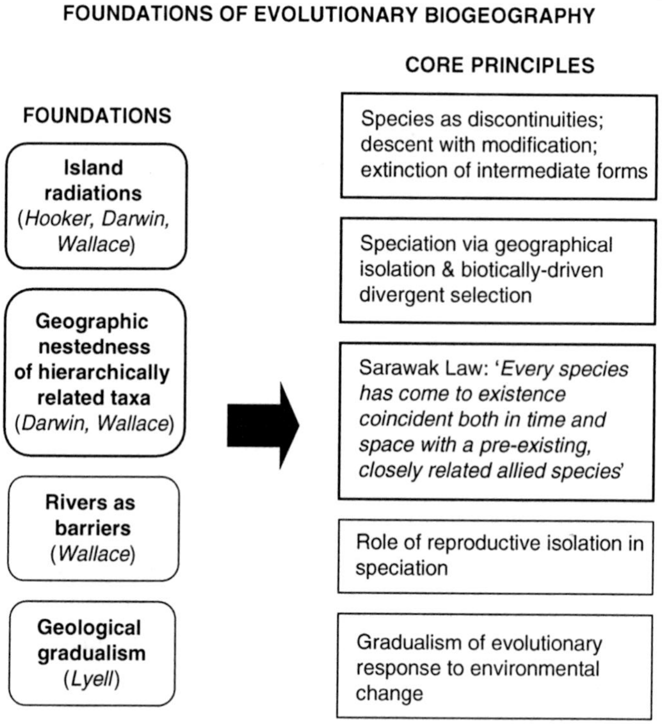

FOUNDATIONS OF EVOLUTIONARY BIOGEOGRAPHY

CORE PRINCIPLES

FOUNDATIONS

Island radiations
(*Hooker, Darwin, Wallace*)

Species as discontinuities; descent with modification; extinction of intermediate forms

Speciation via geographical isolation & biotically-driven divergent selection

Geographic nestedness of hierarchically related taxa
(*Darwin, Wallace*)

Sarawak Law: '*Every species has come to existence coincident both in time and space with a pre-existing, closely related allied species*'

Rivers as barriers
(*Wallace*)

Role of reproductive isolation in speciation

Geological gradualism
(*Lyell*)

Gradualism of evolutionary response to environmental change

FIGURE 7.1 Foundations and core principles of evolutionary biogeography as established by Darwin and Wallace. Topics in bold are discussed in the text.

Our emerging comprehension of the ways in which environmental variation interacts with evolutionary processes to shape diversity can – and should – inform strategies for conservation worldwide.

To achieve this goal we must act on two fronts. First, we must emphasise that biodiversity is dynamic in space and time, and thus, seek to protect the capacity for continuing evolution, as well as the products of past evolution (Frankel, 1974; Moritz, 2002). Second, given sparse information on the distribution of genetic and species diversity, we must better predict how diversity is distributed across regional and local landscapes and how global change will affect such patterns, as well as their underpinning processes (Whittaker *et al.*, 2005). This can be particularly relevant to megadiverse, but poorly documented and highly threatened, tropical regions across the globe.

In this essay, we briefly review the core concepts of evolution established by Darwin and Wallace as they relate to biogeography, and highlight examples where these concepts have been reinforced through the use of current technology. We then proceed to discuss how evolutionary biogeography can evolve from a science in which detailed studies of individual lineages are used to establish principles, to one that seeks to understand how whole assemblages came to be, and to predict biodiversity responses to ongoing, rapid environmental change. In essence, we argue for broadly comparative, integrative and landscape-centric studies of how evolutionary processes intersect with environmental change to generate and sustain diversity.

Reinforced foundations

Darwin and Wallace saw biogeographical patterns as both products and drivers of evolution – an insight lying at the crux of modern-day studies of biodiversity and conservation. Both authors were particularly impressed by island biotas and their apparent ability to diversify after rare, long-distance colonisation from continental sources – often through 'adaptive radiation'. As noted by Darwin, this pattern was more prominent in dispersal-limited taxa, and on remote oceanic rather than continental archipelagos (Gillespie and Roderick, 2002; Whittaker and Fernandez-Palacios, 2007). Examples that substantiate their conclusions abound. Ecological, morphological and molecular studies of the Galapagos fauna demonstrate that species' radiations unique to these remote, volcanic islands typically arise from a limited number of colonisation events from adjacent South America or other regions of the New World (Parent *et al.*, 2008). Similarly, recent molecular studies of the Hawaiian and French Polynesian archipelagos reinforce and extend our knowledge of spectacular radiations from a few common ancestors (Gillespie *et al.*, 2008a; 2008b; Givnish *et al.*, 2009). Whereas most lineages appear to have diversified within the age range of islands currently in the archipelago (<5 million years), a few, such as the extraordinarily diverse Hawaiian fruitflies and lobelias, appear to pre-date the oldest extant islands (Givnish *et al.*, 2009; Price and Clague, 2002). These exceptional cases suggest initial colonisation and diversification on now submerged sea mounts – relating to another fascination of Darwin's: the formation of coral atolls (Darwin, 1842) – with progressive colonisation and diversification on younger islands (the 'progression rule' (Funk and Wagner, 1995)). Interestingly, whereas Darwin largely attributed the evolution of island diversity to divergent selection arising from competition, much of the recent literature emphasises 'ecological release'. This is the idea that lineages diversify post-colonisation to fill open niche space (Gillespie and Baldwin, 2009) – quite different from the intense interactions among species envisaged by Darwin. That said, studies demonstrating parallel development of ecomorphs in island communities (Gillespie, 2004; Losos, 2007; Reding *et al.*, 2009), host-shifts in phytophagous insects (Roderick and Percy, in press), and the interplay between resource competition and phenotypic divergence (Grant and Grant, 1996; Lovette *et al.*, 2002) attest to the importance of strong biotic interactions in shaping adaptive radiations in islands.

Both Darwin and Wallace were strongly influenced by the observation that related groups of organisms tend to be geographically close and nested at higher taxonomic levels (Figure 7.1). That is, species (and higher taxonomic categories) are not randomly distributed on the planet; rather, their geographic position tends to reflect their relationships. This, they reasoned correctly, could only come about through descent with modification (also see Sarawak Law, Figure 7.1). There are, however, striking exceptions that Darwin discussed at length. One is the existence of closely related groups of cool-adapted organisms separated by long distances and climatically divergent zones (Darwin, 1859, Chapter 11). Darwin reasoned that species ranges had been greatly modified under colder climates of past glacial periods, which allowed taxa to cross exposed land bridges between North America and east Eurasia, or expand southwards across currently tropical Central America to colonise temperate environments in the southern hemisphere. His hypothesis was borne out by modern molecular studies that both resolve the relationships amongst cool-adapted, spatially disjunct lineages, and provide estimates of timing for a few key events. Two examples will suffice. Lupins (*Lupinus*) are a diverse group of mostly temperate legumes, with about 100 species in western North America and 85+ in the high Andes. Recent molecular studies (Hughes and Eastwood, 2006) provide strong evidence for colonisation from North America into the Andes in the early–mid Pleistocene (1–2 million years ago), after which the founding lineage diversified rapidly. Similarly, molecular and palaeontological studies support recent evolutionary connections (in geological time) across the Bering land bridge that connected Eurasia and north-western America during glacial periods (Waltari *et al.*, 2007). More generally, Darwin's view that species ranges were profoundly modified during glacial periods has been amply supported by molecular (and palaeontological) analyses of spatial patterns of diversity within temperate species from both North America and Europe. Such phylogeographic studies (Avise, 2004) often reveal high, geographically structured diversity at lower latitudes (reflecting relative stability of populations) and low diversity and signatures of population growth at high latitudes, consistent with range expansions following retraction of icesheets (Hewitt, 2004).

Another emerging pattern with which Darwin and his contemporaries struggled was the existence of related groups of organisms across far-flung, high latitude areas of the southern hemisphere. At first sight, this conflicted with the general argument that related lineages should be geographically proximate, leading to suggestions of historical land bridges or long distance dispersal via ice floes or similar. But Darwin, despite numerous experiments on dispersal ability, remained unconvinced. The explanation for this pattern, of course, is continental drift leading to progressive dismemberment of Gondwana – a concept suggested decades later (Wegener, 1915) and amply supported by subsequent geological and molecular evidence. Yet increasing evidence suggests that both common ancestry in Gondwana and subsequent dispersal have contributed to the current affinities among southern continents (Goldberg *et al.*, 2008; Sanmartín and Ronquist, 2004).

Many authors contributed to the subsequent development of the core evolutionary processes outlined by Darwin and Wallace. In particular, the twentieth-century 'neo-Darwinian' synthesis elaborated the theory of natural selection in a framework of genetic inheritance combined with population processes (Crow, 2008; Provine, 2004). In contrast with some of Darwin's perspectives, some architects of the neo-Darwinian view focused on the importance of rapid, chance-dominated evolution in small isolated populations (Provine, 2004). Darwin, however, was quite adamant that divergence occurs most readily through biotically driven natural selection, especially that due to interspecific competition (Figure 7.1): mere migration into new environments was thought insufficient to generate diversity. Subsequent studies suggest that Darwin and Wallace largely had it right in focusing on divergent natural selection as the engine of diversification, especially when combined with geographic isolation or another mechanism able to suppress genetic exchange (Gavrilets, 2003; Kirkpatrick and Ravigné, 2002; Schluter, 2001). While it remains theoretically plausible that strong 'genetic drift' or 'founder events' cause speciation on islands, or via sexual selection (Uyeda et al., 2009), the evidence points more strongly to divergent selection as a driving force, with or without accompanying founder effects (Clegg et al., 2002; Coyne et al., 1997). One important insight to emerge in the mid–late twentieth century is that evolutionary change in ecologically relevant phenotypes can occur rapidly in natural populations (Carroll et al., 2007); this contrasts with Darwin's view that such change is typically slow.

The challenge: human–driven acceleration of global change

Whilst the field of evolution expanded after the days of Darwin and Wallace, so did the impact of mankind on the same natural systems that they so carefully described and struggled to comprehend. Humans have long affected natural systems, both transforming ecosystems (e.g. by fire; Bowman et al., 2009) and causing or exacerbating extinction (e.g. megafauna; Koch and Barnosky, 2006; island endemics; Steadman and Martin, 2003). Yet both the rate and scale of human impact on nature have increased sharply and dangerously over the past century, particularly the last 50 years. Three processes are of special note: species introductions, habitat conversion and global climate change. Each one has serious consequences on its own – and together, they act synergistically to magnify their individual effects.

Among these processes, climate change is certainly not new to the Earth's biota. Human-induced climate change is nevertheless worrisome because it is leading the planet towards a state that has not existed for millions of years, and this at an exceptionally rapid rate. Biological impacts of such environmental modifications are likely to be of great magnitude at high latitudes, but not exclusively. Contrasting with the initial view that tropical biotas would be relatively unaffected by global warming, recent studies point to great concern for tropical diversity. Novel and disappearing climatic conditions are prevalent in the tropics, especially in montane regions (Williams *et al.*, 2007). Tropical species are thought to have narrower physiological tolerances (Deutsch *et al.*, 2008; Janzen, 1967), and lowland species may be required to move across several degrees of latitude and longitude – often across highly fragmented landscapes – to be able to find climatically matched conditions (Colwell *et al.*, 2008). Thus, the tropics that so amazed Darwin and Wallace, and where the Earth's most species-rich and unique biomes reside, pose us with a double challenge. Not only do we need to predict how and where diversity is distributed, but we must also be able to forecast what subset of biodiversity will be most susceptible to accelerating global change.

The response: biodiversity prediction

Efforts to map current regional or global spatial patterns of species diversity and endemism have been widely implemented in the recent decades. Often, these are led by non-governmental organisations (NGOs) who need to prioritise areas for investment or management. A plethora of approaches and data sources have been used to identify such 'biodiversity hotspots' or priority areas: a few focused on plants, others on birds; some prioritised intact (e.g. wilderness) areas, others concentrated on areas under greater human impact (Whittaker *et al.*, 2005). While there is substantial geographic overlap in the resulting maps (Brooks *et al.*, 2006), all of these approaches take an essentially static view of diversity and are conducted at a coarse geographic scale. Such analyses effectively pinpoint nations most in need of resources to sustain their large share of global diversity. Yet they fail to inform these biodiverse nations on how best to implement conservation actions *within* their jurisdictions. For this purpose, we need to identify geographic foci of diversity within hotspots at a scale relevant to land use planning (Ferrier, 2002). Inevitably, species distribution data become sparser at finer spatial scales, and better known groups (e.g. birds) often fail to predict conservation priorities for other taxa (Moritz *et al.*, 2001).

Two general approaches have been used to overcome these issues and to generate maps of biodiversity distribution at regional or local (biome-specific) scales. One gathers point-locality information from as many species as possible so as to analyse observed distribution patterns, or predictive models of species distributions derived from these data (e.g. Kremen *et al.*, 2008 for Madagascar). Although admirable in scope and effort, these attempts remain incomplete as large numbers of new, often range-limited species are discovered. For example, subsequent work in Madagascar increased the known diversity of amphibian species by 150–200% (Vieites *et al.*, 2009). A second approach, often used by macro-ecologists, seeks to predict spatial trends in species diversity from current environmental data (e.g. area, primary productivity). Yet such models often have substantial error, especially for species with small geographic ranges – the range-restricted endemics for which conservation planning is especially important (Jetz and Rahbek, 2002; Rahbek *et al.*, 2007). This error is likely to be common for low dispersal taxa, in which species diversity is not at equilibrium with current environmental conditions, but rather reflects the imprint of history, especially late Quaternary climate change (see below).

An evolutionary biogeographic approach towards biodiversity prediction

Our goal here is to outline a process-based approach for improving prediction of biodiversity patterns and dynamics across species-rich regional landscapes. To achieve this, we need to understand two non-exclusive categories of processes: those that generate diversity, and those that maintain it subsequently (Figure 7.2). This scheme emphasises adaptive divergence, due to either environmental heterogeneity (i.e. abiotic factors) or strong biotic interactions as observed in hybrid zones (Arnold, 1997) or novel communities (Thompson, 2009), which is most likely to generate new species given some level of geographic isolation

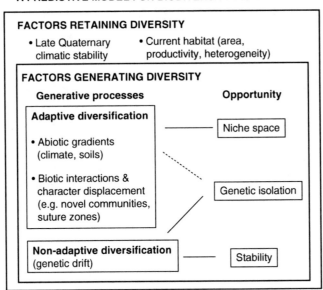

FIGURE 7.2 A synopsis of key processes that generate and sustain species diversity across landscapes (modified from Davis *et al.*, 2008).

and ecological opportunity for persistence. Non-adaptive processes, specifically genetic drift, require complete isolation and, with some exceptions (e.g. under sexual selection; Uyeda *et al.*, 2009), seem less probable causes of speciation over a limited timespan (Orr and Orr, 1996).

There have been many studies of particular systems that document how species diversify in relation to environmental change and these lead to some useful general insights. At large spatial scales, for instance, it has been proposed that the consistently higher species diversity in the tropics reflects a higher propensity for geographic isolation in topographically complex regions because tropical species have narrower and evolutionarily conserved physiological limits (Ghalambor *et al.*, 2006; Janzen, 1967). Analyses of particular tropical lineages, combining phylogenetic hypotheses and distribution data, are shedding light on the dynamics of species divergence across elevational zones (e.g. neotropical salamanders; Kozak and Wiens, 2007; Wiens *et al.*, 2007) and on how evolutionarily constrained physiology can shape biodiversity patterns from global to regional scales (Donoghue, 2008). Along the same lines, recently evolved species ('neo-endemics') are often seen to cluster in regions with steep environmental gradients, novel environments, or recent climatic stability (Davis *et al.*, 2008; Fjeldså and Lovett, 1997). Multi-taxa comparisons within the Amazon have revealed that current patterns of species and genetic diversity stem from historical (tectonic arches) more than current (riverine) barriers to gene flow (Gascon *et al.*, 2000; Patton *et al.*, 2000). Islands provide another example where knowledge of environmental history improves biodiversity prediction: factors such as the dynamics of island formation, topographic complexity and area, and connectivity during periods of lower sea level significantly improve our understanding of spatial patterns of island endemism (Heaney, 2007; Whittaker *et al.*, 2008). Comprehending the mechanisms by which speciation processes play out across landscapes remains a significant and important challenge. Molecular analyses, when combined with information on trait evolution and landscape history, have much to offer in this regard (Moritz *et al.*, 2000).

Subsequent to speciation, the persistence and geographic extent of species may be affected by a wealth of factors, such as environmental changes, competition, invasive species and pathogens. Yet, as Darwin correctly anticipated, the first – fluctuations in climatic conditions across the landscape – appears paramount. Recent studies clearly demonstrate that climate-driven glacial and interglacial periods of the Pleistocene (particularly the last glacial maximum), as well as climatic fluctuations of smaller amplitude during the Holocene, had a dominant influence on current spatial patterns of diversity (Hewitt, 2004; Jansson and Dynesius, 2002). Although current environmental parameters – habitat area, heterogeneity, net primary productivity, etc. – explain much variation in local-scale diversity, regional species and genetic diversity bear a strong imprint of late Quaternary climatic history, especially for low dispersal species (Graham *et al.*, 2006).

In the tropics, temperatures cooled significantly at the last glacial maximum. Yet there was considerable spatial variation in the magnitude of cooling, and even greater variability has been detected in the effects on precipitation or seasonality. Areas that remained cool but sufficiently wet to sustain rainforest, such as the western Amazon and Guiana Shield, had greatly modified communities as some currently mid-montane species mixed with lowland taxa (Colinvaux *et al.*, 2000; Rull, 2005), and in such areas, it appears that populations of some forest-dependent species remained large (Lessa *et al.*, 2003). The evolutionary impact of this dramatic and repeated reshuffling of rainforest communities is unknown. By contrast, where rainfall or temperatures decreased to the extent that tropical rainforests retracted to local refugia (e.g. mountains or adjacent wet coastal plains), the effects were profound. In such circumstances, spatial modelling of the effects of climate-driven fluctuations in the distributions of rainforest habitats and species provide powerful predictions about biodiversity patterns. These, in turn, can be tested through comparative analyses of genetic diversity. Where these models are verified, we expect that they should effectively predict patterns of local endemism in low dispersal organisms – likely the bulk of locally unique diversity within hotspots.

Description and examples

Our approach to process-based biodiversity prediction relies on the spatially explicit modelling of biological responses to former climate change. From geographic information systems (GIS) based models of species (or habitat) distribution under current and past climates, we develop spatially explicit predictions about patterns of genetic diversity and historical demography, which can then be tested with molecular data (Figure 7.3). Where possible, the GIS models should also be validated by direct fossil evidence (Martínez-Meyer *et al.*, 2004; Petit *et al.*, 2002; Svenning *et al.*, 2008), though such records are sparse for much of the tropics. The tools needed for this approach – including point records spanning species' ranges (Graham *et al.*, 2004), GIS layers for current habitats and regional palaeoclimates (PMIP2; http://pmip2.lsce.ipsl.fr), and statistical methods for inferring past demography based on DNA sequences (Hey and Machado, 2003; Hickerson *et al.*, 2006; Lemmon and Lemmon, 2008; Nielsen and Beaumont, 2009; Richards *et al.*, 2007) – have only recently become available and continue to improve. These GIS models make many assumptions (e.g. species limits are set by climate and

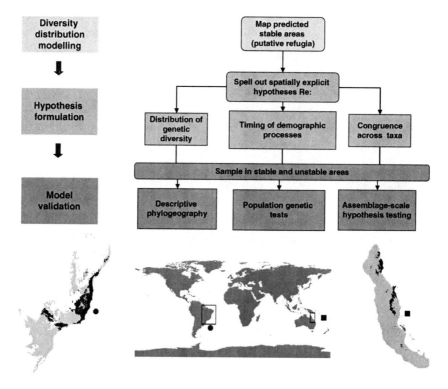

AN EVOLUTIONARY BIOGEOGRAPHIC APPROACH TO BIODIVERSITY PREDICTION IN HOTSPOT AREAS

FIGURE 7.3 An approach to predicting the impact of late Quaternary climate change on the current distribution of diversity within biodiversity hotspots. At the bottom are two examples of modelled refugia, which are predicted to contain a high proportion of unique diversity (left, Brazilian Atlantic rainforest; right, Australian wet tropics).

are invariant, etc.), but we should keep in mind that they are hypotheses subject to testing, not the answer in themselves.

The essence of the approach is as follows. Based on detailed information about spatial distribution of habitat (maps of structural forest types) or species occurrence (georeferenced point records), one first builds correlative models of the habitat or species of interest under current climate conditions. These estimate the climatic conditions within which species or habitats currently occur. These climate conditions are then projected onto palaeoclimate surfaces to retrodict past distributions of the species or habitat in question. By contrasting distribution patterns and ranges across different time periods, we can then predict the location(s) of historically stable 'refugial' areas (areas persistently occupied by the species or biome of interest), as well as relatively unstable areas. Such spatially explicit predictions naturally lead to demographic hypotheses of past range contractions, range expansions or population isolation, which can be tested with molecular genetic data from multiple species.

We illustrate the approach with two case studies: the Australian wet tropics (AWT) rainforest of north-eastern Queensland, and the Atlantic rainforest (AR) in Brazil. Both are located across geologically ancient montane systems and adjacent coastal plains, and the AWT system has a pronounced dry season. For the AWT, a series of palaeomodelling exercises (Nix, 1991; VanDerWal *et al.*, 2009) predict that rainforests and most widespread rainforest species retracted to geographically isolated coastal mountains and adjacent lowlands during the cold, dry conditions of the last glacial maximum. By contrast, specialists of montane rainforests are predicted to be most sensitive to warm, wet conditions as prevailed in the mid Holocene (Moussalli *et al.*, 2009).

The spatial predictions of stable rainforest areas in the AWT: (1) improved prediction of patterns of species richness for low dispersal groups (but not more dispersive birds) (Graham *et al.*, 2006); (2) correctly predicted spatial patterns of genetic diversity in snails (Hugall *et al.*, 2002), beetles (Bell *et al.*, 2007), and most amphibians and reptiles (Moussalli *et al.*, 2009; Schneider *et al.*, 1998); and (3) revealed how past changes in rainforest distribution led to the formation of current multispecies hybrid zones (Moritz *et al.*, 2009). Climate driven rainforest fluctuations have shaped diversity such that much of the unique genetic and species diversity is now concentrated on the coastal mountains that sustained rainforest through the late Pleistocene. Of great concern, many of these areas are predicted to be most strongly impacted by future climate warming (Williams *et al.*, 2003).

Applying this predictive approach to the Brazilian Atlantic rainforest yielded new insights and predictions regarding the distribution of diversity within this highly imperilled biodiversity hotspot. Modelling of (pre-clearing) rainforest under current and past climates predicted greater late Quaternary stability of the forest in the northern section relative to the south (Figure 7.3). This is significant because the northern forests have received much less attention from biologists and are more impacted by recent habitat loss. These models are generally congruent with previously reported areas of endemism and fossil pollen records (Carnaval and Moritz, 2008). Similar results were also obtained for individual species of frogs that occur in low to mid altitudes. In this case, the modelling of palaeodistributions correctly predicted observed patterns of genetic diversity (higher in northern than southern areas), reflecting long-term local persistence in the north and recent range expansion to the south (Carnaval *et al.*, 2009). Following from these analyses, we predict that the northern lowland forests contain higher genetic diversity and more range-restricted species than currently known. Surveys are now under way to test this prediction, and the results will inform regional efforts to protect diversity in the face of continuing land use pressure and climate change.

Building on the Darwin/Wallace legacy

We hope to have made clear the power and importance of combining knowledge of evolutionary processes with biogeographic analysis – an exercise built on the foundations so ably laid by Darwin and Wallace in the mid nineteenth century. Modern-day science has the means, the opportunity and the responsibility to guide conservation of global biodiversity in the face of human-driven environmental change. Biodiversity prediction is central to this task, and will guide research and management of natural areas. Our recent studies are demonstrating the value of modelling processes that enabled tropical diversity to be maintained in space and time, particularly through investigations of biological responses to late Quaternary climate change. Despite the fact that not all tropical areas have been subject to such strong climate-driven fluctuations in rainforest area, evidence from non-equatorial forests in South America and Australia argue that palaeomodelling offers an effective method for improving biodiversity prediction within such hotspots, and perhaps also in other climatically analogous systems (e.g. Western Ghats and Sri Lanka; Bossuyt *et al.*, 2004; Kahindo *et al.*, 2007 for East Africa montane forests). Methods for modelling persistence under past climate change will no doubt improve, and we have much to learn about how to predict speciation dynamics across different types of landscapes. None the less, there is a need and an opportunity now to apply knowledge of these general processes with information on regional environmental history to improve knowledge of diversity patterns and dynamics in other poorly known lineages and regions.

Acknowledgements

We thank Rosie Gillespie, Sonal Singhal and Roberta Damasceno for insightful comments and discussions. Funding has been provided by the National Science Foundation (DEB 0817035 to C. M., DBI 0512013 to A. C. C.).

8 Postgenomic Darwinism

JOHN DUPRÉ

Introduction

I might perhaps have better called this chapter post-Darwinian genomics. One point I want to make is that it is time we disconnected our
discussions of evolution from an unhealthily close connection with the
name of Charles Darwin. Darwin, after all, wrote his most famous work
150 years ago, and rapidly advancing sciences do not generally rest
directly on work a century and a half old. Darwin knew nothing of
genetics or genomics and, as I shall especially emphasise, there have also
been remarkable advances in microbiology that he could not have known
about and that fundamentally affect our understanding of evolution.

I do not, of course, have any wish to deny Darwin's greatness as a
scientist. It is impossible to read his extensive scientific writings without
being struck by the powers of his observation, the encyclopaedic breadth
of his knowledge, and a remarkable ability to move between detailed
observation and the grand sweep of theory. Moreover, the fact that it
was Darwin who convinced the learned world of the fact of evolution,
of the common descent of humans and other forms of life, gives him an
uncontestable place in the history of ideas. This has provided a corner-
stone of the naturalistic world view which, if hardly the universal
perspective of the human race, has increasingly become the dominant
perspective among its most educated and reflective minorities.

But this is not just a quibble about an anomalous degree of deference
to a distinguished and influential dead scientist. I think this deference

Darwin, eds. William Brown and Andrew C. Fabian. Published by Cambridge University Press. © Darwin College 2010.

can act as an obstacle to the advance of the science. At its most extreme –
and here one cannot help seeing an ironic defeat for biology in its debate
with religious creationists – Darwin takes on the role of scriptural
authority, and his words are subject to detailed exegetical analysis as if
this was a way to better understand the biological world. It sometimes
seems that Darwin, like God in war, appears on both sides of most major
biological debates. One of the great epistemic virtues of science is that it
constantly attempts to revise itself and advance its understanding as new
information or insight accumulates. Excessive deference or even reverence for past authorities is the antithesis of this epistemic commitment.

But more subtle and specific misunderstandings are also associated
with the excessive reverence for Darwin. It is sometimes forgotten that
whereas, after the publication of the *Origin of Species,* Darwin quite
rapidly convinced the learned world of the truth of evolution, the transformation between distinct species, conviction for the process commem-
orated in his subtitle, natural selection, was not achieved widely until
well into the twentieth century, with the synthesis of Darwinian natural
selection with Mendelian genetics. The real target of this chapter is not
concerned so much with Darwin's own views, but with the view that
emerged at that time as the 'New Synthesis', and has evolved today into
what is often called neo-Darwinism. There is a popular view that Darwin
got just about everything right that was possible for someone deprived
of an adequate understanding of genetics, and the New Synthesis filled in
this final gap. And it is this vision, lent weight by the towering authority
of Charles Darwin, which I suggest is becoming an obstacle to the
advancement of our understanding of evolution and its ability to take
account of the very remarkable advances in our biological understanding
over the last few decades.

Neo-Darwinism

By 'neo-Darwinism' I mean the New Synthesis as modified by the
emergence of molecular genetics in the 1950s and beyond. From the New
Synthesis it maintains (in addition to the core commitment to natural
selection) the Mendelian idea of inheritance as particulate, the concept
of genes that are transmitted to offspring in their entirety or not at all, and

the concept, following August Weismann, of a sharp division between germ cells, which carry the transmitted genes, and somatic cells. Neo-Darwinism can be defined, for my present purposes, in terms of two core theses and one important corollary. The first thesis is that overwhelmingly the most important cause of the adaptation of organisms to their environment, or conditions of life, is natural selection. This is the heart of the Darwinism in neo-Darwinism. The second thesis is that inheritance, at least as far as it is relevant to evolution, is exclusively mediated by nuclear DNA. This thesis could be seen, if a little simplistically, as a blend of Mendel and Weismann seen through the lens of Crick and Watson.

The corollary, especially stemming from the Weismannian ingredient of the second thesis, is the rejection of Lamarckism. Lamarckism here has perhaps less to do with the actual opinions of Jean-Baptiste Pierre Antoine de Monet, Chevalier de la Marck even than do contemporary understandings of Darwinism with the ideas of Charles Darwin. Lamarckism now has come to mean the inheritance, or bequeathal to descendants, of somatic characteristics acquired in the lifetime of an organism, and this has become the ultimate taboo in Darwinian theory. The significance of the taboo is that it presents a powerful restriction on the variations that can be the targets of natural selection, the differences between which Nature selects. These differences are now assumed to be, or to be direct causal consequences of, randomly generated changes in the genes or genome of the organism.

In the following pages I shall describe some developments in recent biology that show that neo-Darwinism, if not entirely obsolete, is at least severely limited in its ability to encompass the full range of evolutionary processes. My suggestion is that the association with a long-dead hero can convey the message that in general outline the problems around evolution have been solved long ago, and only the details, perhaps of evolutionary history, need to be sorted out. This message is sometimes explicitly promoted in opposition, particularly in the United States, to the powerful voices of creationists opposed to the very idea of evolution. As I have already suggested, the response is surely a counterproductive one. We should celebrate the fact that the exploration of evolution is an exciting scientific project and, far from being essentially complete, it is one of which we are still only at the very early stages. Those who insist on having the whole 'truth' whether or not we have any serious grounds for believing it are perhaps closer to the religious fundamentalists they so vehemently oppose than they would like to believe. At any rate, what I shall do in the main body of this essay is look at some areas of biological research that are radically altering our views of evolution and challenging neo-Darwinian orthodoxy. As should by now be clear, I take this as illustration of the excitement and dynamism of evolutionary science, certainly not any indication of its vulnerability.

Revisionist Darwinism 1: the tree of life

The first topic I want to address that will indicate the shakiness of the neo-Darwinian orthodoxy is the concept of the tree of life. The tree of life is the standard neo-Darwinian representation of the relatedness of organisms. As a tree, crucially, it constantly branches, and branches always diverge, never merge. Species are represented as small twigs; larger branches represent larger groups of organisms. By following down from the branches towards the trunk of the tree it is possible in principle to work backwards through all the ancestors of a group of organisms to the earliest beginnings of life at the tree's base. Darwin's imprimatur for this divergent evolutionary structure is often secured by a picture in the notebooks that seems to represent a divergently branching structure, accompanied, to the delight of philosophical commentators, by the legend 'I think' (Figure 8.1). More significant still, though, is the sole illustration in the *Origin of Species* representing with a branching diagram the formation of new species through the divergence of varieties within a species, an illustration that follows a chapter adumbrating the benefits of divergence by analogy with the division of labour (Figure 8.2).

But this image of the tree of life has been rendered at least partially obsolete by recent developments, especially in microbiology, where so-called lateral gene transfer, the passage of genetic material not from ancestors, but from sometimes distantly related organisms on widely separated branches of the tree of life, is common. One reason for the importance of this phenomenon is that it threatens to undermine the pattern of explanation of features of biological organisms that is universally mandated by the divergently branching structure of the tree.

John Dupré

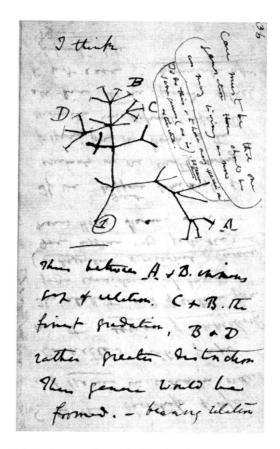

FIGURE 8.1 Darwin's first sketch of an evolutionary tree from
Notebook [B], the first notebook on *Transmutation of Species* (1837). Reproduced by
kind permission of the Syndics of Cambridge University Library.

Neo-Darwinism, it will be recalled, attributes the adaptation of organisms to natural selection, working on variations in the genetic material.
These variations are generated endogenously and transmitted within the
narrow confines of the species, understood as groups of organisms
sharing access to the same gene pool. Embedding this idea within the
wider frame of the tree of life, we can see that the explanations for all the
characteristics of an organism are to be sought in the sequence of
ancestors traceable down the branches of the tree, and in the evolutionary process, namely natural selection, to which these ancestors had been

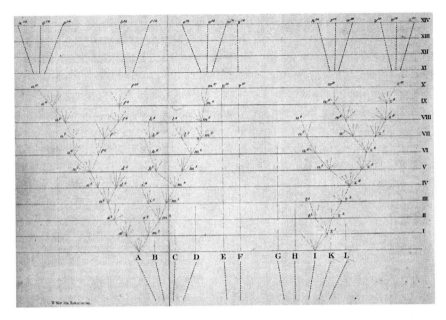

F I G U R E 8 . 2 The sole illustration from *The Origin of Species*, showing the divergence of ancestral species, first into varieties and eventually species. Species G and H, for instance, have gone extinct, whereas species I eventually gives rise to six descendant species. Reproduced by kind permission of the Syndics of Cambridge University Library.

subject. Explanation of the characteristics of an organism by lateral gene transfer, on the other hand, puts no limit in principle on where in the history of life a particular aspect of a lineage may have originated. This is immediately obvious when we note that if lateral gene transfer is common, the overall structure of relations between organisms will take the form not of a tree, but of a web, or net. And in a web, unlike a tree, there are many paths from one point to another.

Lateral gene transfer is widely recognised to be endemic among microbial life forms (see, for example, Doolittle, 1999). Microbes transfer bits of DNA from one to another by a process sometimes likened to sex called *conjugation*, in which a tube down which the genetic unit passes is inserted by one cell into another; by *transformation*, the uptake of free DNA from the environment; and by *transduction*, in which the transfer is mediated by viruses. These processes can result in genetic transfers between the most distantly related organisms, even organisms from different domains, the threefold classification now taken to be the most fundamental division of living organisms.[1] This, in short, removes the presupposition that the evolutionary exigencies of linear ancestors explain the features of their living descendants. Lateral gene transfer allows features to have come from, more or less, anywhere in the biosphere.

Questioning the tree of life remains, none the less, a controversial business.[2] Although many microbiologists have accepted that there is no unique tree for microbes, some still resist this conclusion, and insist that there is a core genome, resistant to lateral transfer, and in terms of which a microbial phylogenetic tree can be reconstructed (Lawrence and Hendrickson, 2005; but see Charlebois and Doolittle, 2004). There are serious problems with this, however. First we might wonder, even if the claim can be sustained in some sense, whether the tree based on the core genome is very useful. Or in other words, what is the tree of life for? If, as I have been suggesting, its function is to underpin evolutionary explanations of organismic features, then the more prevalent is lateral transfer, the less will any tree be able to serve this end. This is even more so as the genes that are likely to form the constant core will inevitably be ones with fundamental, and therefore substantially invariant, functions across a very wide range of organisms. They will, for that reason, be the least useful in tracking differences between organisms. This leads naturally to the question, why track phylogeny using these genes rather than some others. Because of lateral transfer not all trees of genes will coincide. And it may be that different gene trees will be useful for answering different questions. Perhaps the defenders of the core genome have in mind that what they should attempt to construct is the cell tree, the tree that traces the sequence of (vertical) cell divisions back to the beginning of cellular life. The trouble then is that this seems just to assume what is at issue, that vertical inheritance is what really matters. If this position is to be maintained regardless of how much the contents of the cells may be changed by other interacting, non-vertical processes, one might wonder in the end whether it would end up as little more than a fetishism of the cell membrane.

Eukaryote[3] biologists are generally much more confident of the tree of life, and with good reason.[4] Lateral gene transfer seems less common among eukaryotes, and there is little question that the tracing of vertical ancestral relations is a powerful and useful way of classifying these organisms.[5] Even here, though, there is reason to be cautious. For a start, hybridisation seems to be much more common than was once thought (Mallet, 2008). But perhaps more important, the transfer of genetic elements by viruses certainly does continue in eukaryotes, and may well prove to be an important factor in evolution. About half of the human genome, for instance, consists of material that is thought to have originated in transfers from viruses. Much of this, it is true, consists of highly repetitive sequences that have seemed unlikely to be functionally significant. When the idea of 'junk DNA' was fashionable, these were prime candidates for junk. However, it now appears that at least 70% of the genome is transcribed into RNA, and investigation of the roles of various kinds of RNA fragment in regulating the genome is one of the fastest growing fields in molecular biology. It would be premature to assume that sequences of viral origin may not play crucial roles in such regulatory systems. And finally, there are examples of significant functional features of cells that do appear to involve protein-coding sequences of viral origin. The best example here is of the evolution of placental mammals. The tissue that provides the barrier between fetal and maternal circulations, the syncytium, is believed to be coded for by genes of viral origin (Mallet *et al.*, 2004). There may surely be other equally significant cases. It is at any rate clear that, even among eukaryotes, lateral origins play some role in explaining the current features of organisms. The always branching, never merging tree of traditional phylogeny is not enough.

Revisionist Darwinism 2: evolution by merger

Lateral gene transfer can be seen in a rather different light as an example of something much broader: evolution by merger. This gets to one of the most general points I want to make about the limitations of neo-Darwinism. The first thesis mentioned above, the overwhelming emphasis on natural selection, has encouraged neo-Darwinian evolutionists to think a great deal about competition, but very little about cooperation. Indeed, the latter appears mainly in the guise of a problem – the 'problem of altruism'. The problem of altruism is, crudely put, the problem of understanding why it is that, in a 'Darwinian' world in which the only survivors are the most ruthless and self-interested competitors, some organisms are actually nice to one another. But looked at from a quite different perspective, life is a massively cooperative enterprise and 'altruism' should hardly be surprising. The elements in a cell or the cells in a multicellular organism must obviously work in a highly coordinated way and subordinate their own 'interests' to those of the whole of which they are part. It will be objected at once that this is cooperation within an organism, not between organisms, and so of course not a problem. But this reply assumes that we know exactly what constitutes an organism and what is merely a part of an organism, an assumption I shall suggest is highly problematic.

It is perhaps hardly controversial to note that natural selection will frequently select the organisms that are best at cooperating with the organisms with which they interact. This is just one way of adapting to the environment, the most salient part of which is typically the other organisms that inhabit it. I want to go a step beyond this, however, and suggest that merger with other organisms (or suborganismic biological entities) is a central process by which biological organisms evolve. One such process is referred to as endosymbiosis, and is most widely familiar from the ideas of Lynn Margulis (1970) about the origins of the eukaryotic cell.[6] It is now universally acknowledged that the mitochondria that provide the energy source for all eukaryotic cells, and the chloroplasts that effect photosynthesis in plants, were both originally free-living organisms but are now more or less independently reproducing but wholly dependent constituents of larger cells. Although the details are much more controversial, it is also believed by many that the eukaryotic cell itself derived from a merger between two prokaryotes, perhaps a bacterium and an archaeon.

The examples just mentioned are instances of fully obligate endosymbiosis: mitochondria are parts of eukaryotic cells, and there is no more question of why they are acting altruistically towards the containing cell than of why my liver acts altruistically towards me. However, it is important to note that endosymbiosis is something that may evolve over a long period of time, and in the mean time may consist of a range of degrees of interdependence from conditional and reciprocal cooperation to full endosymbiosis. There are, for example, well studied cases of varying degrees of endosymbiosis between insects and bacteria. *Buchnera aphidicola*, endosymbionts of aphids, have been associated with their partners for up to 200 million years, and have lost the ability to carry out various essential metabolic functions on their own. *Wolbachia*, on the other hand, a genus of bacteria associated with a very wide range of arthropod species including perhaps half of all insect species, is generally referred to as a parasite. *Wolbachia* are particularly interesting for their ability to control the reproductive behaviour of their hosts. Some can kill or feminise males, or induce parthenogenesis. They can also induce reproductive incompatibility between insects infected with different *Wolbachia* strains, possibly playing a determinant role in speciation.

It is generally supposed that the manipulation by *Wolbachia* of their hosts' reproduction contributes to their own rather than their hosts' reproductive interest. However, as some host species appear unable to reproduce without the assistance of *Wolbachia*, and as *Wolbachia* are obligatorily symbiotic, it is not always clear how these interests are to be separated. *Wolbachia* are involved in transfers of DNA between insect species, raising questions about genetic differentiation of insect species (Whitworth *et al.*, 2007), and a whole *Wolbachia* genome has been found embedded within a *Drosophila* genome (Dunning Hotopp *et al.*, 2007). It has also been found that *Wolbachia* may reduce the vulnerability of their hosts to viral infections (Teixeira *et al.*, 2008). It would be difficult to assess the ratio of costs and benefits to the parties in these intimate associations, but it seems likely that this balance will vary from case to case, and that in some cases the relationship has moved to full mutualism or even symbiosis.

One reason I have spent a little time on this example is that it begins to introduce a fundamental question, namely how we determine the limits of an organism. No one doubts that mitochondria are parts of the organisms in which they are found whereas, on the whole, everyone takes *Wolbachia* and their insect hosts to be distinct organisms. But what is the basis of this different treatment? It will be recalled that discussions of altruism tend to assume that this question is unproblematic. If, as I shall suggest, it is a thoroughly indeterminate matter, settled as much by our interests as investigators as by anything in Nature, it will clearly be necessary to rethink the question of altruism or, more broadly, competition and cooperation.

John Dupré

What is an organism?[7]

Although philosophers have for many years questioned some of the key concepts of biology, such as the species or, more recently, the gene, on the whole they have not seen much to worry about with the concept of an organism. According to the orthodox view, there are two kinds of organisms: single-celled, or microbes, and multicellular, or (as I have elsewhere suggested we call them (Dupré and O'Malley, 2007)) macrobes.[8] In the former case the cell is the organism. In the latter case all the cells derived from a fertilised egg, or zygote, constitute the one organism. We might summarise the view as 'one organism, one genome'. This concept of the organism could be seen as the microlevel reflection of the macroscopic tree of life: both within and between organisms we find orderly and always divergent branching. But we might also want to approach the question of what constitutes an organism from a functional perspective: what are the systems of cells that interact with the surrounding environment as organised and generally cooperative wholes? From this starting point we would note that microbes do not typically function as isolated individuals but rather in complex associations often composed of highly diverse kinds of cells. Typical of such associations are biofilms, the generally slimy coatings that develop on practically any moist surface. Consider, for instance, one well studied class of biofilms, those on the surfaces of our teeth known as dental plaque. Over 500 different bacterial taxa have been found living in the human mouth (Kolenbrander, 2000) and, according to one authority, 'Oral bacteria in plaque do not exist as independent entities but function as a coordinated, spatially organized and fully metabolically integrated microbial community, the properties of which are greater than the sum of the component species' (Marsh, 2004). Why would we not consider this community, the organised functional whole, to constitute an organism?

If we concede that biofilms comprise a kind of multicellular organism, then the argument is also over as far as traditional monogenomic multicellular organisms are concerned. For all known such multicellular wholes exist in symbiotic relations to often enormous and diverse communities of microbes. In the human body, for instance, it is estimated that 90% of the total number of cells are in fact microbial (Savage, 1977), living mainly in our gut, but also on the surface of the skin and in all the bodily orifices. These microbes contain perhaps 100 times as many genes as those found in the more traditional human genome (Xu and Gordon, 2003), which has led to the launch by the US National Institutes of Health of the Human Microbiome Project, which will explore this missing 99% of the full human genome. The importance attached to this project reflects an increasing awareness that these symbiotic microbes have a fundamental influence on human health. They are known to be involved in digestive processes, and hypothesised to have a significant role in causing obesity. For model organisms it has been demonstrated that microbial symbionts are necessary for normal physiological development (Bates *et al.*, 2006), that they affect gene expression in the 'host' cells (Hooper *et al.*, 2001), and that they are involved in the maturing of the immune system (Umesaki and Setoyama, 2000).[9] There is every reason to expect similar findings in humans.

I propose then that the typical organism is a collection of cells of different kinds, organised cooperatively to maintain its structure and reproduce similar structures. As Maureen O'Malley and I have put it (2009), an organism is a metabolically integrated community of lineage segments. It will immediately strike evolutionists that this conceptually separates the organism (functional whole) from the evolving entity (part of a lineage). But this, of course, is the point. The assimilation of these concepts obscures the empirical reality that evolution requires both (directly) reproducing lineages and the assembly of organisms from components of these lineages, and that these are in principle quite independent processes. While most of these lineage segments will have little chance of reproducing themselves except in so far as they are able to form parts of appropriate communities, this is nevertheless a contingent matter.[10] One consequence of this proposal is that what is an organism, and whether something is part of an organism or not, are not questions that necessarily admit of definitive answers. Whether a group of microbes is a closely connected ecological community or an organism may be a matter of biological judgement. The important point is that it, or most of it, will share an evolutionary fate. If its constituent cells are to send descendants off to participate in new biofilms it will be because the parental biofilm is thriving. What I have been calling organisms are units of selection, objects between which natural selection selects.

Cooperation again

I can imagine a frustrated reader complaining that I have yet to address the kind of cooperation that is of real interest to evolutionary biology, cooperation between conspecifics. Cooperation with other organisms is just adaptation to the environment, of which they are part. Some of them are to be eaten, most can be ignored, others are more useful as collaborators, and so on. Conspecifics, on the other hand, are always competitors for representation by their descendants in subsequent generations. So let me say something about this topic.

The orthodox neo-Darwinian view is that the only circumstance that brings about cooperation between conspecifics is kin selection. Here it is time to distinguish between some degrees of cooperation. If two lions can kill a wildebeest that neither could handle alone, and moreover it will provide plenty of food for both, they will do well to cooperate. Evolutionists tend rather to speak of 'altruism' in a technical sense according to which an act is altruistic only if it not only confers a benefit on the recipient but is also more costly to the donor than refraining from action. Any animal that acted in this way would lose out to natural selection in competition with others that avoided such acts of kindness. The only exception would be the case where the beneficiary is kin, perhaps one's offspring, as described by so-called inclusive fitness theory. Here the fundamental principle is said to be Hamilton's rule: $rB > C$. B is the benefit to the recipient, C the cost to the donor and r the so-called coefficient of relatedness. This coefficient is ½ for offspring or full siblings in sexual species, and is thought of as the proportion of genes that two organisms share by virtue of their relations of descent.[11] If $rB > C$, for example if I make a sacrifice that provides more than double the benefit to my child, evolution will favour such behaviour. I don't want to deny that this is a powerful tool for analysing important aspects of evolutionary processes and their potential stability or instability. One very impressive example is its application to theoretical discussions of the evolution of eusociality, the often vast and complexly articulated social system characteristic of many ants, wasps and bees (Hymenoptera), termites and, alone among mammals, the naked mole rat. These arguments have shown that only under conditions of strict monogamy for an exclusive breeding couple is such a social arrangement likely to evolve. Recent work (Hughes *et al.*, 2008) has confirmed that such strict monogamy was indeed the ancestral condition in a large number of Hymenoptera species studied, giving convincing support to inclusive fitness theory.

I want to make two somewhat more sceptical comments on this topic, however. First, it is often said that altruism outside the narrow confines of kin selection theory will be subverted by competition from less altruistic rivals. The assumption that there are indeed such rivals seems sometimes to be a matter of pure dogma. Consider, for instance, an example that seems to contradict standard kin selection theory, also from the Hymenoptera. The Argentine ant (*Linepithema humile*), while known for its inter-colony aggression in its native land, has now taken to behaving in a non-aggressive, cooperative way in relation to other colonies of conspecifics, in a range of newly colonised areas in Europe, North America, Japan and Australia. Contrary to a speculation that this must be due to genetic relatedness between the recently landed colonists, colonies in the European case, at least, were found to be genetically diverse. There is considerable dispute about how to explain or even describe this phenomenon, though one thing that seems to be widely agreed is that the ants as a whole do very well out of the arrangement. As humans have also discovered, warfare may benefit a few, but it is hardly good for the species. Unsurprisingly, it is also speculated that the arrangement will be unstable. A mutant aggressive colony would perhaps do extremely well cutting a swathe through its amiable neighbours. But even if this could happen, it doesn't imply that it must. Perhaps eventually the system will collapse, and perhaps it is bound to do so in the very long run. But, to paraphrase Keynes, in the long run we are all extinct. The existence of cooperation between non-kin is sufficient to show that there are evolutionary processes capable of creating it. The most widely discussed such process is of course group selection, though this does remain controversial, if less so after the extremely influential work of Sober and Wilson (1998). Even if it is demonstrated that there are circumstances that would undermine these cooperative systems, this hardly shows that they could not, after all, have come into being in the first place. It is a contingent matter how long more or less cooperative, even altruistic, systems last.

This brings me conveniently to my second point, the one that has been the main focus of this paper. It is that arguments about what entities can be expected to cooperate or compete with what others presuppose that we know what the individuals are that are cooperating or competing. Group selection is taken to be problematic because it is assumed that the members of the group are real, robust, indisputable individuals, whereas we see the group itself as a fragile coalition, a thoroughly dubious individual. But what I have been suggesting is that in fact there is no sharp line between the group of more or less cooperative individuals on the one hand, and the unified self-contained individual on the other. Indeed, it may well be that there is a tendency for the former to evolve into the latter, and that in the course of this process the individuals will act increasingly as parts subordinate to a larger whole. Presumably something like this must have happened in the evolution of multicellularity, and indeed is thought to have happened many times (Buss, 1987).

There is even a bigger picture here. The idea that life is hierarchically structured is an ancient and obvious one. Molecules comprise cells; cells make organs and organ systems; organisms are composed of organs and the like; and organisms in turn make up larger social or ecological units. This is a useful picture in focusing the investigating mind on particular aspects of the biological world, but it can easily be taken too literally. Cells, organs, and even organisms are, in Nature, embedded in larger systems, and their separate existence requires either a scalpel or a process of abstraction. Two further points reinforce both the significance and the plausibility of this observation. First, or so I would also argue, a full understanding of a biological entity at any of these intermediate levels is impossible without taking account both of its composition from smaller constituents, and the influences exerted on it by the larger system of which it is part, though that is an argument beyond the scope of the present paper (see Powell and Dupré, 2009). Causal explanation runs both from smaller to larger and from larger to smaller. Second, we should recall that our hierarchy of entities is already itself an abstraction from a hierarchy of processes. It may be that many forms of scientific reasoning require descriptions of entities as if they had a set of static properties definitive of such entities. But the reality, as best we understand it, is of a series of nested processes at timescales ranging from nanoseconds for intercellular chemical reactions to hundreds of millions of years for some macroevolutionary processes (Dupré, 2008). The illusion of an objectively distinct and unique hierarchy of objects is much less compelling when this abstraction is borne in mind.

Lamarck redux

I turn now to the strictest taboo in neo-Darwinism, Lamarckism.[12] Lamarckism, here, must be understood in an even less historically grounded sense than Darwinism, and has little to do with the great French naturalist. The taboo concerns the inheritance of characteristics acquired during the lifetime of the organism. According to strict neo-Darwinists only genetic mutations within the germline and the recombination of genetic resources brought about by sexual reproduction provide the resources on which selection acts. Curiously, however, though mention of Lamarckism can still bring a shudder to many evolutionary biologists, almost no one still believes in the strict form of the taboo. Or so, anyhow, I shall attempt to demonstrate.

The topic with which I began, lateral gene transfer, is one generally acknowledged qualification of strict anti-Lamarckism. Genes transferred laterally into the genome of an organism are certainly acquired, and may certainly be inherited. The reason that Lamarckism is such a profound potential challenge to traditional Darwinism is that somatic traits acquired during the lifetime of the organism may often be adaptive, constituting the organism's response to the environment. An animal may run as fast as it can to escape speedy predators or in pursuit of fleet-footed prey, for example, and in doing so it may develop stronger leg muscles. But the inheritance of such adaptive acquired characteristics would threaten the first principle of neo-Darwinism, the monopoly of natural selection in producing adaptation.[13] Here it may be thought that lateral gene transfer offers little threat of this kind. Perhaps we should see it as no more than the equivalent of a very big mutation. But first, there is a growing consensus that lateral gene transfer has been of fundamental importance, at least in microbial evolution. Boucher and colleagues (2003) review the evidence for its role in 'photosynthesis, aerobic respiration, nitrogen fixation, sulfate reduction, methylotrophy, isoprenoid biosynthesis, quorum sensing, flotation (gas vesicles), thermophily, and halophily'. Moreover, second, a large number of researchers suggest that lateral gene transfer is indeed often an adaptive response to the environment. According to Pal and colleagues (2005) 'bacterial metabolic networks evolve by direct uptake of peripheral reactions in response to changed environments'. And 'lateral gene transfer provides the bacterial genome with a new set of genes that help it to explore and adapt to new ecological niches' (Marri *et al.*, 2007). Note the similarity with the kind of cooperative ventures I discussed earlier in this paper. Whole microbial cells (or indeed macrobial cell systems) adapt to their environment by recruiting, or being recruited by, coalitions of cooperating cells. More complex organisms may recruit conspecifics or even members of other species to form social collectives that enhance their ability to cope with environmental challenges. And, finally, cells may sometimes recruit adaptively useful genetic fragments from their environments. All very Lamarckian.

One response to the issue of lateral gene transfer may be to downplay the importance of microbial evolution. Perhaps microbes are really rather insignificant little beasts? To this, however, it is sufficient to respond that 80% of evolutionary history is a history solely of the evolution of microbes; the vast majority of organisms alive today are microbes; and all known macrobes are dependent for their existence on symbiotic relations with microbes. As I have briefly mentioned above, the importance of lateral gene transfer in macrobial evolution is itself a matter of active debate. But anyhow, an account of evolution that doesn't apply to microbes is one that ignores the overwhelmingly dominant manifestation of life on Earth.

Varieties of inheritance

The Lamarckian aspects of the topic just considered at least do not violate the idea that the vast majority of inheritance passes through the nuclear genome. Lateral gene transfer may be very important in evolution, but it is very rare by comparison to the routine passage of genetic material from parents to offspring. However, there are other reasons to recognise that the neo-Darwinian restriction of inheritance to transmission of the nuclear genome provides a thoroughly impoverished picture.[14] The most widely discussed form of inheritance that is excluded is cultural inheritance. Much of this discussion is directed specifically to human evolution (e.g. Richerson and Boyd, 2005). Although this work is very important in many ways, including in showing the inadequacy of the orthodox neo-Darwinian treatments of human evolution offered by evolutionary psychologists, in the present essay I shan't discuss the special problems of human evolution. There is still heated debate about whether human evolution raises unique issues, and every aspect of human evolution has been discussed and debated by numerous authors, including myself (Dupré, 2001). In this essay I shall avoid these very specific issues.

I mentioned in passing above the perspective of developmental systems theory (DST) (see note 10). DST abandons the myopic focus on the nuclear genome typical of much neo-Darwinism, and looks at the entire cycle of events by which the organism is reproduced. The fundamental unit of analysis is the life cycle of the organism and, given this unit of analysis, it should be clear from the preceding discussion that the requisite concept of an organism must also be the multigenomic, multilineage one advocated above. From a DST perspective a large body of work on the cultural transmission of behaviour can be seen as fitting fully into an evolutionary framework. Some fairly arbitrarily selected recent examples are the learning of frog calls by bats (Page and Ryan, 2006), the use of sponges in foraging by bottle-nosed dolphins (Krutzen et al., 2005), or, perhaps the best studied example, the transmission of bird songs (Slater, 1986). The process of learning behaviour by immature individuals, and the behaviour of mature individuals involved in mating and in rearing offspring, are clearly crucial parts of the developmental cycle, and potentially evolving aspects of the life cycle.

Less familiar, but perhaps even more important, is the fact that far from the idea occasionally suggested in popularisations of neo-Darwinism (e.g. Dawkins, 1976) that the genome is the only significant material thing transmitted in reproduction, the minimal material contribution in any form of reproduction is an entire maternal cell. This is an extremely complex object with a great deal of internal structure and a bewildering variety of chemical constituents. For asexual organisms (most organisms, that is), it seems perverse to think of anything other than the cell as the basic unit of inheritance. For sexual organisms the issue is more complex, because each individual begins life with a new, generally unprecedented, inheritance, at least genetically. But of course there is a vast number of other materials that are passed on with the maternal cell (and a few even with the paternal sperm) that form a major part of the (inherited) developmental system.

It is sometimes supposed that all the non-DNA material passed on in reproduction is unimportant because it is the DNA that carries the inherited differences on which natural selection can act. But this seems to be a dogmatic assertion rather than anything for which there is empirical evidence. Why, for example, might not changes in the chemistry of the cell membrane be inherited in the process of cell division? But we do not need to speculate. There is a rapidly developing field of biological research, epigenetics, which may be seen as answering a fundamental question, but one that can seem mysterious from the radically DNA-centred perspective – why do different cells with the same genome do different things? Why do my liver cells differ so radically from my brain cells, for instance? Central to epigenetic research is the understanding of how other chemicals in the cell act on the genome to determine which parts of it are expressed (i.e. transcribed to RNA and (sometimes) translated to a protein).

Epigenetics is important in part for breaking the hold of the so-called 'central dogma' of molecular genetics, that causality, and hence information, runs only in one direction, from DNA to RNA to protein.[15] Epigenetics could be described, with a little hyperbole, as the study of the falsity of the central dogma. But, secondarily and consequently, it reveals the potential diversity of inheritance at the molecular level. In the first place, once it is seen that the surrounding cell acts on the genome, not merely the other way around, it is clear that the inter-generational transmission of any part of the cellular system may embody significant heredity. Second, one of the crucial ways in which epigenetic effects on the DNA occur is through actual modifications to the structure of the DNA chain. The best studied of these is methylation, in which a methyl group is attached to one of the bases, cytosine, that comprise the DNA sequence. This has the effect of inhibiting the transcription of the sequence of which the methylated cytosine molecule is part. It is an obvious possibility that these modifications could be inherited. The claim that they are indeed inherited has been highly controversial. In part this was because it had been understood that a process of demethylation took place during meiosis, the formation of sex cells. If this demethylation was total, then the epigenetic changes would not be transmitted. Recently, it has become increasingly widely agreed that demethylation is not complete, and hence that methylation is to some degree inherited (Chong and Whitelaw, 2004). This has been a remarkably heated controversy, and it is impossible to avoid the suspicion that this is in significant part because if methylation patterns, something that can be acquired in the lifetime of an organism, can be inherited, this will raise the possibility of violating the taboo against Lamarckian inheritance.

It is very interesting to note that epigenetic changes might still be inherited even if they had proved to be entirely erased at meiosis. This is because when they are induced by external, environmental influences they may also contribute to the production of those same influences. The classic example substantiating this possibility derives from a series of experiments on maternal care in rats, carried out by Michael Meaney and colleagues. Grooming, especially licking, by mother rats appears to be very important for the proper development of rat pups, and rats that do not receive sufficient such maternal care grow up generally fearful and, most significantly, less disposed to provide high quality maternal care to their offspring (Weaver *et al.*, 2004; Meaney *et al.*, 2007). It has been demonstrated that these effects are mediated by maternal grooming causing changes of methylation within cells in the brain, which in turn affect the production of neurotransmitters. Thus, the trait of high quality maternal care appears to be transmitted through the induction of methylation patterns in young female rats through exposure to such maternal care. This might also be seen as an adaptive and heritable epigenetic switch: in a stressful and dangerous environment, perhaps, it is best to be fearful (even the paranoid can be right) and too risky to devote more than the minimum effort to caring for the young. It is, of course, possible, and a possibility that might be very widely significant, that this modestly Lamarckian mechanism could be an adaptation acquired by Darwinian means. As mentioned above, inheritance mechanisms are among the more interesting features of organisms that evolve.

Conclusion

I conclude very much as I began. With absolutely no disrespect to Darwin, biological insights gained over the last few decades have profoundly altered the way we can and should think about evolution. It appears that evolutionary processes may be more diverse than we had imagined, including Lamarckian mechanisms as well as neo-Darwinian, cooperative and symbiotic as well as competitive and individualistic.[16] The evolutionary histories of the entities that make up biological wholes may also be multiple. Genomes have different histories from the organisms in which they reside, both because they assimilate material from other sources, and because they have their own history within the organism – for example, of intragenomic duplications. And organisms, at least when understood as the functional wholes that interact with the rest of the world, are coalitions of entities with diverse evolutionary histories. Neo-Darwinism has much to say about the divergent processes that push biological entities apart, much less about the convergent processes in which the whole is constantly more than the sum of the parts.

None of this should be remotely shocking. But for some reason or reasons we have buffered an outdated view of evolution with a thicket of surrounding dogma and presumption that stands in the way of advancing the theory in line with the stunning insights that are being gained in other parts of biology. Part of this story surely is that this dogma has developed as an unintended response to competition with thoroughly anti-scientific perspectives (creationism, 'intelligent design') that have somehow positioned themselves as rivals to scientific evolutionism. And I suspect the links with creationist views may be more complex than that. Extreme neo-Darwinists sometimes share with creationists the yearning for an all-encompassing scheme, a single explanatory framework that makes sense of life.[17] One thinks, for instance, of Daniel Dennett's (1995) paean of praise for natural selection, which he then deploys as the essential resource to explain everything from the breeding behaviour of bees to the deliberative processes of the human mind. But evolution is a mosaic of more or less related processes, producing a motley collection of outcomes. Just because one has a hammer, one should be careful not to suppose that everything is a nail.

If one of the things that needs to be done to remedy this partial paralysis of our evolutionary thinking is that we detach our view of evolution a little from our reverence for Charles Darwin, then I am sure he won't mind.

Acknowledgements

Many of the ideas in this essay derive from joint work with Maureen O'Malley, though I don't assume she will agree with everything I have written. It has also been much improved by detailed comments on an earlier draft by her and by Staffan Müller-Wille. I gratefully acknowledge the support of the UK Economic and Social Research Council (ESRC). The research in this paper was part of the programme of the ESRC Centre for Genomics in Society (Egenis).

Notes

CHAPTER 1

1 Antony Hippisley Coxe (1973), *Haunted Britain; a Guide to Supernatural Sites Frequented by Ghosts, Witches, Poltergeists and other Mysterious Beings.* Hutchinson. With photos taken by Robert Estall, p. 54. This essay draws on work published as 'Making Darwin: biography and changing representations of Charles Darwin' (Browne, 2010).

2 From a wide range of literature on the relations between fact and fiction in modern biography, see O'Connor (1991) and Holdroyd (2002).

3 See Shortland and Yeo (1996) and Söderqvist (2007). After an early call for attention from Henry Guerlac (1954), there is now increasing interest in meta-biographical issues. See, for instance, Rupke (2005) *Alexander von Humboldt: A Metabiography* and Higgitt (2007) *Recreating Newton: Newtonian Biography and the Making of Nineteenth-century History of Science.* In relation to imagery of scientists, see Lawrence and Shapin (1998) *Science Incarnate: Historical Embodiments of Natural Knowledge* and Frayling (2005) *Mad, Bad and Dangerous? The Scientist and the Cinema.*

4 This estimate is based on a fairly strict definition of the biographical genre. In addition, there are a very large number of studies of Darwin's theories, either historical or explanatory in scope, and many studies on evolution in general that include aspects of his life story. A conservative estimate derived from the combined library catalogue Copac indicates at least 2000 books in this larger category since 1882.

5 A useful survey of Darwin biographies is given by Bowler (1990). For an analysis of conceptual shifts in Darwin scholarship, see Bohlin (1991).

6 Browne (2003); see also 'Presidential Address: Commemorating Darwin,' *British Journal for the History of Science*, **38**, 2005, 251–74; and 'Looking at Darwin: portraits and the making of an icon', *Isis*, **100**, 2009, 542–70, by Browne.

7 The original manuscript is in Cambridge University Library, DAR 26. This was published in part by Darwin's son Francis (Darwin, F. 1887). Later it was freshly transcribed and published in full by Nora Barlow in 1958 with the original omissions restored.

8 Classic studies include Manier (1978); Howard E. Gruber, *Darwin on Man: A Psychological Study of Scientific Creativity,* together with Darwin's early and unpublished notebooks, transcribed and annotated by Paul H. Barrett. Foreword by Jean Piaget (New York: E. P. Dutton), 1974.

9 On Darwin as a collector, see Porter (1985) and Browne (1995). Alfred Russel Wallace noted this similarity in tastes and attributed much of his and Darwin's theoretical achievement to a mutual engagement with traditional collecting and classifying pursuits; see Alfred Russel Wallace, 'Address' in *The Darwin–Wallace celebration held on Thursday 1st July 1908* (London: Linnean Society of London, 1908).

10 Burkhardt and Smith *et al.* (1985–). A synopsis of each item of correspondence is online at http://darwinproject.ac.uk. See also J. Secord (1991).

11 See Browne (2006) and Browne (2002). Lyell's impact on geology is discussed in the introduction by J. Secord (1998).

12 See note 10.

13 The Victorian commitment to Baconian science is discussed in the introductory essay to Hull (1973). See also Yeo (1985).

14 Browne (2003). For portraits in science, see Jordanova (2000b).

15 Smiles (1859); see also Anne Secord (2003) and Travers (1987). More general reflections of Victorian moral imagery can be found in Collini (1991).

16 Atkins (1974). After an international fundraising effort, Down House was purchased and given to the nation by the Wellcome Trust in 1996. I gratefully acknowledge English Heritage, especially Down House curator Tori Reeve, and former members of the English Heritage Down House Committee charged with restoring and refurbishing for reopening the house in 1998, especially Randal and Stephen Keynes.

17 The purchase and possible use of the house were not self-evident in the 1920s. Papers from the Evolution Committee of the Royal Society of London indicate that before the First World War, Down House was on Andrew Carnegie's shopping list, with a proposal to turn it into an international research station. Next came Henry Fairfield Osborn, the palaeontologist and director of the American Museum of

Natural History, who offered American funds to establish the house as a centre for evolutionary research. Osborn had given Darwin the grand finale in his history of biology, *From the Greeks to Darwin*, published in 1895. But Osborn's advocacy ran against the views of the Royal Society's Evolution Committee, headed by William Bateson, and the plan foundered. 'Is Darwin the right man for evolutionists to remember', asked one member of the committee – a question that would be inconceivable today. I am grateful to Marsha Richmond for information that Thomas Hunt Morgan, the experimental geneticist of Cornell University, offered some of his Nobel Prize money to help establish a scientific research station at Down. These various hopes were expressed in different ways during the 1909 commemorations in London and Cambridge of the centenary of Darwin's birth.

18 Mayr (1982; 1999). On Mayr's role as a historian of biology, see especially Winsor (2006).
19 Keith (1955). Keith is quoted from Colp Jr (1989), p. 172.
20 See also Browne 2002, note 11, pp. 114–25.
21 See the introduction to Barrett *et al.* (1987). The long version of the *Origin of Species* was transcribed and published as Stauffer (1975).
22 Cambridge University Library, Darwin Archive, 'D' notebook, foliated as D135e, dated 28 September 1838, transcribed in Barrett *et al.* (1987), p. 375.
23 Adler (1959), on which see also Browne (1990a, b). The many interpretations of Darwin's illnesses are discussed by Colp (1977) and Fleming (1961).
24 Browne (1980). Also discussed by Kohn (1985) and Ospovat (1981).
25 A rough indication of this shift in perspective is seen in the number of pages in recently published Darwin biographies that cover the period of his life after the *Origin* was published (some 23 years). Bowlby (1990) gives this period one-quarter of the total number of pages. Desmond and Moore (1991) give it one-third. Browne (2002) gives it an entire volume.

CHAPTER 2

1 C. Darwin to J. D. Hooker, 23 September 1864. In Burkhardt, F. *et al.* (eds) (2001). *Correspondence of Charles Darwin.* Cambridge: Cambridge University Press, **12**, pp. 336–7.
2 See special section on 'Darwin as cultural icon' in *Isis*, **100**, 2009.
3 *The Times of India*, 22 April 1882.

4 C. Darwin to J. D. Hooker, 11 January 1844. In Burkhardt, F. *et al.* (eds) (1987). *Correspondence of Charles Darwin.* Cambridge: Cambridge University Press, **3**, pp. 1–3.

5 Darwin, C. (2008). *Recollections of the Development of My Mind and Character.* In *Evolutionary Writings*, ed. C. Darwin. Oxford: Oxford University Press, p. 411. On Huxley, see Desmond, A. (1997). *Huxley: From Devil's Disciple to Evolution's High Priest.* London: Penguin.

6 On the early history of Darwinism, see Moore, J. R. (1991). Deconstructing Darwinism: the politics of evolution in the 1860s. *Journal of the History of Biology,* **24**, 353–408.

7 The Coming Man. *The World,* 12 March 1871. A copy is in Cambridge University Library, Add. mss DAR 140.1.1.

8 *The Press,* 13 June 1863. On Butler's essays, see Amigoni, D. (2007). *Colonies, Cults and Evolution: Literature, Science and Culture in Nineteenth-century Writing.* Cambridge: Cambridge University Press, pp. 142–63.

9 Quoted in Pusey, J. R. (1983). *China and Charles Darwin.* Cambridge, MA: Harvard University Press, p. 185.

10 Quoted in Xiaobing Tang (1996). *Global Space and the Nationalist Discourse of Modernity: The Historical Thinking of Liang Qichao.* Stanford: Stanford University Press, p. 50 and 57.

11 See the passages translated by M. Elshakry in Darwin, C. (2008). *Evolutionary Writings.* Oxford: Oxford University Press, pp. 229–30.

12 Compare the work translated in Keddie, N. R. (1983) as *An Islamic Response to Imperialism: Political and Religious Writings of Sayyid Jamāl ad-Dīn ʻal-Afghāni'.* Berkeley and Los Angeles: University of California Press, with later comments translated by M. S. Elshakry, in Darwin, C. (2008) *Evolutionary Writings,* ed. J. Secord. Oxford: Oxford University Press, pp. 431–2.

13 See http://darwin-online.org.uk for the printed works and many manuscripts; http://www.darwinproject.ac.uk for the letters.

14 For helpful discussions of this and the issue of use-inheritance, see Endersby, J. (2009). An evolving *Origin.* In *On the Origin of Species,* ed. C. Darwin. Cambridge: Cambridge University Press, pp. 377–95.

15 C. Darwin to J. D. Hooker, 29 March 1863. In Burkhardt, F. *et al.* (eds) (1999). *Correspondence of Charles Darwin.* Cambridge: Cambridge University Press, **11**, pp. 277–9.

CHAPTER 4

1 Toulouse School of Economics (IDEI/GREMAQ). This is an edited version of a lecture delivered in the 2009 series of Darwin Lectures

at Darwin College, Cambridge, to mark the 200th anniversary of Darwin's birth and the 150th anniversary of the publication of *The Origin of Species.* I am grateful to the Master and Fellows of Darwin College and Alexey Titarenko for permission to reproduce one of Alexey's remarkable photographs.

2 The literature on primate societies is vast; an excellent place to start is de Waal (2001).

3 World Health Organization, Mortality Database.

4 Letter to J. D. Hooker, 13 July 1856, Darwin Correspondence Project, http://www.darwinproject.ac.uk/darwinletters/calendar/entry-1924. html#back-mark-1924.f2.

5 Roughgarden (2009) chastises Darwin's theory of sexual selection for placing too much emphasis on competition and not enough on cooperation in evolution. She is right about how widespread cooperation is throughout the animal kingdom, as well as about the wonderful variety of forms it takes. But I believe she underestimates how much Darwin himself was aware of this, and also how much cooperation in some dimensions is compatible with cooperation in others.

6 This does not mean that their interests are opposed; they have some interests in common and some that diverge.

7 Evolution gems, no. 14, *Nature,* January 2009.

8 These examples are just some among many surveyed in Arnqvist and Rowe (2005).

CHAPTER 8

1 These three domains, probably still not widely familiar to lay readers, are the bacteria, the archaea and the eukarya. Most people are still probably more familiar with the outdated fivefold classification: animalia, plantae, fungi, protista and monera. But in fact the first four of these are now understood to belong within just one of the currently distinguished fundamental domains, the eukarya. The remaining kingdom, monera, is now divided between the domains archaea and bacteria (and some miscellaneous organisms are now seen as eikarya). Evidently this change reflects a vastly increased awareness of the diversity of microbes, which include the latter two domains as well as the protista and many of the fungi (yeasts). As will be emphasised below, microbial life is the dominant form of life on Earth in terms of its antiquity, its diversity and even its sheer mass.

2 One reason for this briefly alluded to above is that, especially in the US, any dissent from Darwinian orthodoxy can be seen as providing

ammunition for creationists. I have occasionally had the dubious pleasure of finding my own work cited with approval on creationist websites. As also noted above, however, for science to insist on orthodoxy in defence against theological fundamentalism could provide only the most Pyrrhic victory.

3 Eukaryotes, it will be recalled, are the animals, plants and fungi, as well as a diverse and miscellaneous collection of microbes. They are distinguished by a more complex cell structure which includes a distinct compartment, the nucleus, which houses the genetic material.

4 As an important qualification, there is no consensus about the evolutionary origins of eukaryotes.

5 Notice that the tendency to lateral gene transfer, or mechanisms to prevent it, are contingently evolved features of particular classes of organisms, so there is no reason why its prevalence among microbes should imply anything about multicellular eukaryotes.

6 Margulis and Sagan (2002) present a broader argument for the central importance to evolution of what I am calling evolution by merger.

7 This section summarises work in a number of recent papers, including O'Malley and Dupré (2007) and Dupré and O'Malley (2007; forthcoming).

8 It is bizarre that we should have a term covering the vast majority of organisms that have ever existed, but none for the small minority to which this term does not apply. 'Macrobe' seems the obvious candidate to fill this gap; and filling it should help to deflect the illusion that it is microbes that are the exceptional or unusual biological forms.

9 For a general review of this topic, see McFall-Ngai (2002).

10 This proposal is very much in the spirit of developmental systems theory (Oyama, 2000; Oyama et al., 2001), though developmental systems theorists have paid too little attention, to date, to the place of symbiotic microbes in developmental systems.

11 This is a trickier notion than is often perceived, as can somewhat amusingly be noticed when the same authors suggest that we share 50% of our genes with our siblings and 98% with a chimpanzee. Needless to say, different conceptions of 'the same genes' are at work here (see Barnes and Dupré, 2008, p. 98). This shouldn't cause any confusion in the present context, however.

12 The only openly Lamarckian work that has been widely influential is Jablonka and Lamb (1995), who review in detail the Lamarckian implications of the epigenetic inheritance that will be briefly described in the next section.

13 A very important book by Mary Jane West-Eberhard (2003) explores a growing body of findings on the remarkable ways in which the development of organisms responds to environmental circumstances, and the implications of this for evolution. West Eberhard refuses to call any of her views Lamarckian, though this may possibly be for fear that this would lead to their marginalisation.

14 It is useful here to remember that inheritance systems of any kind are products of evolution. Perhaps some very primitive kind of inheritance must have emerged at the dawn of evolution, but whatever inheritance systems exist today were certainly products of evolution and not, as some presentations of neo-Darwinism sometimes suggest, a prerequisite. In this light it will hardly be surprising if several have evolved.

15 In its original formulation by Francis Crick, the 'central dogma' referred only to information about the sequence of bases or amino acids in macromolecules. Although even this is no longer a defensible position, in view of phenomena such as alternative splicing, the expression is nowadays typically used much more broadly, and in the present discussion I adopt this broader use. It is also plausible that Crick intended a degree of irony in using the term 'dogma', something that has curiously dissipated in some contemporary references.

16 Here I should note that Darwin, as opposed to the later neo-Darwinians, was always a good pluralist and, incidentally, increasingly a Lamarckian.

17 I am grateful to Staffan Müller-Wille for emphasising this point to me in conversation.

References

REFERENCES (JANET BROWNE – CHAPTER 1)

Abir-Am, P. and Elliott, C. (eds) (1999). *Commemorative Practices in the Sciences: Historical Perspectives on the Politics of Collective Memory.* Osiris no. 14. Chicago: University of Chicago Press.

Adler, S. (1959). Darwin's 'illness'. *Nature*, **184**, 1102–3.

Atkins, H. (1974). *Down, the Home of the Darwins: The Story of a House and the People who Lived There.* London: Royal College of Surgeons of England.

Barlow, N. (ed.) (1933). *Charles Darwin's Diary of the Voyage of H.M.S. Beagle.* Cambridge: Cambridge University Press.

Barlow, N. (1945). *Charles Darwin and the Voyage of the Beagle.* London: Pilot Press Ltd.

Barlow, N. (ed.) (1958). *The Autobiography of Charles Darwin 1809–1882.* With the original omissions restored. London: Collins.

Barlow, N. (1963). Charles Darwin's ornithological notes. *British Museum (Natural History) Bulletin. Historical series*, **2**(7), 201–78.

Barlow, N. (ed.) (1967). *Darwin and Henslow: The Growth of an Idea; Letters, 1831–1860.* London: John Murray [for] Bentham-Moxon Trust.

Barrett, P. H., Gautrey, P. J., Herbert, S., Kohn, D. S. (eds) (1987). *Charles Darwin's Notebooks, 1836–1844: Geology, Transmutation of Species, Metaphysical Enquiries.* Cambridge: Cambridge University Press.

Bettany, G. (1887). *Life of Charles Darwin.* London: Walter Scott, p. 170.

Bohlin, I. (1991). Robert M. Young and Darwin historiography. *Social Studies of Science*, **21**, 597–648.

Bowler, P. J. (1990). *Charles Darwin: The Man and His Influence.* Cambridge, MA: Cambridge University Press, pp. 1–16.

Brackman, A. C. (1980). *A Delicate Arrangement: The Strange Case of Charles Darwin and Alfred Russel Wallace.* New York: Times Books.

Brooks, J. L. (1984). *Just Before the Origin: Alfred Russel Wallace's Theory of Evolution.* New York: Columbia University Press.

Browne, J. (1980). Darwin's botanical arithmetic and the 'principle of divergence', 1854–1858. *Journal of the History of Biology,* **13**, 53–89.

Browne, J. (1990a). Spas and sensibilities: Darwin at Malvern. In *The Medical History of Waters and Spas,* ed. Roy Porter. London: Wellcome Institute for the History of Medicine, pp. 102–13.

Browne, J. (1990b). 'I could have retched all night': Charles Darwin and his body. In *Science Incarnate: Historical Embodiments of Natural Knowledge,* eds C. Lawrence and S. Shapin. Chicago: University of Chicago Press, pp. 240–87.

Browne, J. (1995). *Charles Darwin: Voyaging.* New York: Knopf.

Browne, J. (2002). *Charles Darwin: The Power of Place.* New York: Knopf.

Browne, J. (2003). Charles Darwin as a celebrity. *Science in Context,* **16**, 175–94.

Browne, J. (2006). *On the Origin of Species: A Biography.* London: Grove Atlantic.

Browne, J. (2010). Making Darwin: biography and changing representations of Charles Darwin. *Journal of Interdisciplinary History,* **40**, 347–73.

Burkhardt, F. and Smith, S. (eds) (1985). *A Calendar of the Correspondence of Charles Darwin, 1821–1882.* New York: Garland Publications, Introduction, pp. 1–3.

Burkhardt, F., Smith, S. *et al.* (eds) (1985–). *The Correspondence of Charles Darwin,* 16 vols., 1821–1868. Cambridge: Cambridge University Press.

Cantor, G. (1996). The scientist as hero: public images of Michael Faraday. In *Telling Lives in Science: Essays on Scientific Biography,* eds Shortland, M. and Yeo, R. Cambridge: Cambridge University Press, pp. 171–93.

Churchill, F. B. (1982). Darwin and the historian. In *Charles Darwin: A Commemoration,* ed. R. J. Berry. London: Linnean Society of London, pp. 45–68.

Collini, S. (1991). *Public Moralists: Political Thought and Intellectual Life in Britain, 1850–1930.* Oxford: Oxford University Press.

Colp Jr., R. (1977). *To be an Invalid: The Illness of Charles Darwin.* Chicago: University of Chicago Press.

Colp Jr., R. (1989). Charles Darwin's past and future biographies. *History of Science,* **27**, 167–97.

Coxe, A. H. (1973). *Haunted Britain; a Guide to Supernatural Sites Frequented by Ghosts, Witches, Poltergeists and other Mysterious Beings.* With photos taken by Robert Estall. London: Pan Books, p. 54.

Cubitt, G. and Warren, A. (eds) (2000). *Heroic Reputations and Exemplary Lives.* Manchester: Manchester University Press.

Darwin, F. (ed.) (1887). *The Life and Letters of Charles Darwin, Including an Autobiographical Chapter*, 3 vols. London: John Murray, **Vol. 1**, pp. 26–107.

de Beer, G. (1958). *Charles Darwin*. London: Oxford University Press.

de Beer, G. (ed.) (1960). Darwin's notebooks on transmutation of species. *British Museum (Natural History) Bulletin. Historical series* **2**, nos 2–6: 3.

de Beer, G. (1965). *Charles Darwin: A Scientific Biography*. Garden City, NY: Doubleday, published in cooperation with the American Museum of Natural History.

Desmond, A. J. and Moore, J. R. (2009). *Darwin's Sacred Cause: How a Hatred of Slavery Shaped Darwin's Views on Human Evolution*. London: Michael Joseph.

Eiseley, L. (1958). *Darwin's Century*. New York: Doubleday.

Eiseley, L. (1979). *Darwin and the Mysterious Mr. X: New Light on the Evolutionists*. London: Dent.

Fara, P. (2000). Isaac Newton lived here: sites of memory and scientific heritage. *British Journal for the History of Science*, **33**, 407–26.

Fara, P. (2002). *Newton: The Making of Genius*. New York: Columbia University Press.

Ferry, G. (1998). *Dorothy Hodgkin: A Life*. London: Granta Books.

Fleming, D. (1961). Charles Darwin, the anaesthetic man. *Victorian Studies*, **IV**, 219–36.

Frank, A. J. and James, L. (2005). An 'open clash between science and the church?': Wilberforce, Huxley and Hooker on Darwin at the British Association, Oxford, 1860. In *Science and Beliefs: From Natural Philosophy to Natural Science, 1700–1900*, eds David M. Knight and Matthew Eddy. Aldershot: Ashgate Publishing, pp. 171–93.

Frayling, C. (2005). *Mad, Bad and Dangerous? The Scientist and the Cinema*. London: Reaktion Books.

Friedman, A. J. and Donley, C. C. (1985). *Einstein as Myth and Muse*. Cambridge: Cambridge University Press.

Geikie, A. (1888). Darwin's life and letters. In *Littell's Living Age*, 5th Series, Vol. LXI. Boston: Littell and Company, p. 3.

Geison, G. L. (1995). *The Private Science of Louis Pasteur*. Princeton, NJ: Princeton University Press.

Greene, M. (1993). Recent biographies of Darwin: the complexity of context. *Perspectives on Science*, **1**, 659–75.

Guerlac, H. (1954). Lavoisier and his biographers. *Isis*, **45**, 51–62.

Herbert, S. (2005). *Charles Darwin: Geologist*. Ithaca: Cornell University Press.

Higgitt, R. (2007). *Recreating Newton: Newtonian Biography and the Making of Nineteenth-century History of Science*. London: Pickering and Chatto.

Himmelfarb, G. (1959). *Darwin and his Darwinian Revolution.* London: Chatto and Windus.

Holder, C. F. (1891). *Charles Darwin: His Life and Work.* New York: Putnam, p. vi.

Holdroyd, M. (2002). *Works on Paper: The Craft of Biography and Autobiography.* London: Little Browne and Company.

Holroyd, M. (1994). *Lytton Strachey: The New Biography.* London: Chatto and Windus, pp. 149–50, 269.

Hull, D. (ed.) (1973). *Darwin and his Critics: The Reception of Darwin's Theory of Evolution by the Scientific Community.* Chicago: University of Chicago Press.

Huxley, J., assisted by J. Fisher (1959). *The Living Thoughts of Darwin.* Greenwich, Conn.: Fawcelt Publications, p. 1.

James, F. (2008). The Janus face of modernity: Michael Faraday in the twentieth century. *British Journal for the History of Science,* **41**, 477–516.

Jordanova, L. (2000a). Remembrance of science past. *British Journal for the History of Science,* **33**, 387–406.

Jordanova, L. (2000b). *Defining Features, Scientific and Medical Portraits 1660–2000.* London: Reaktion Books.

Keith, A. (1955). *Darwin Revalued.* London: Watts.

Koerner, L. (2000). *Linnaeus: Nature and Nation.* Cambridge, MA: Harvard University Press.

Kohn, D. (1985). Darwin's principle of divergence as internal dialogue. In *The Darwinian Heritage,* ed. D. Kohn. Princeton: Princeton University Press, pp. 245–57.

Lack, D. (1947). *Darwin's Finches.* New York.

La Vergata, A. (1985). Images of Darwin: a historiographic overview. In *The Darwinian Heritage,* ed. D. Kohn. Princeton: Princeton University Press, pp. 901–72.

Lawrence, C. and Shapin, S. (eds) (1998). *Science Incarnate: Historical Embodiments of Natural Knowledge.* Chicago: University of Chicago Press.

Manier, E. (1978). *The Young Darwin and his Cultural Circle: A Study of Influences which Helped Shape the Language and Logic of the First Drafts of the Theory of Natural Selection.* Boston: D. Reidel Pub. Co.

Mayr, E. (1982). *The Growth of Biological Thought: Diversity, Evolution, and Inheritance.* Cambridge, MA: Belknap Press.

Mayr, E. (1999). *Systematics and the Origin of Species, from the Viewpoint of a Zoologist.* Cambridge, MA: Harvard University Press.

Mayr, E. and Provine, W. B. (eds) (1980). *The Evolutionary Synthesis: Perspectives on the Unification of Biology.* Cambridge, MA: Harvard University Press.

Mellersh, H. E. L. (1964). *Charles Darwin, Pioneer in the Theory of Evolution*. London: A. Barker.

Moore, J. (1996). Metabiographical reflections on Charles Darwin. In *Telling Lives in Science: Essays of Scientific Biography*, eds M. Shortland and R. Yeo. Cambridge: Cambridge University Press, pp. 267–81.

Moorehead, A. (1969). *Darwin and the 'Beagle'*. London: H. Hamilton.

O'Connor, U. (1991). *Biographers and the Art of Biography*. Dublin: Wolfhound Press.

Osborn, H. F. (1924). *Impressions of Great Naturalists: Darwin, Huxley, Balfour, Cope and Others*. New York: Scribner's Sons, p. xiv.

Ospovat, D. (1981). *The Development of Darwin's Theory: Natural History, Natural Theology, and Natural Selection, 1838–1859*. Cambridge: Cambridge University Press.

Pancaldi, G. (2003). *Volta: Science and Culture in the Age of Enlightenment*. Princeton: Princeton University Press.

Pitts, M. E. (1995). *Toward a Dialogue of Understandings; Loren Eiseley and the Critique of Science*. Bethlehem: Associate University Press, pp. 101–43.

Porter, D. (1985). The *Beagle* collector and his collections. In *The Darwinian Heritage*, ed. D. Kohn. Princeton: Princeton University Press.

Rupke, N. (2005). *Alexander von Humboldt: A Metabiography*. Frankfurt am Main and New York: Peter Lang.

Secord, A. (2003). 'Be what you would seem to be': Samuel Smiles, Thomas Edward, and the making of a working-class scientific hero. *Science in Context*, **16**, 147–73.

Secord, J. A. (1991). The discovery of a vocation: Darwin's early geology. *British Journal of the History of Science*, **24**, 133–57.

Secord, J. A. (ed.) (1998). *Charles Lyell, Principles of Geology*. London: Penguin Classics.

Shapin, S. (1991). A scholar and a gentleman: the problematic identity of the scientific practitioner in early modern England. *History of Science*, **29**, 279–327.

Shapin, S. (1994). *A Social History of Truth: Civility and Science in Seventeenth-century England*. Chicago: University of Chicago Press.

Shermer, M. (2002). *In Darwin's Shadow: The Life and Science of Alfred Russel Wallace*. Oxford: Oxford University Press.

Shortland, M. and Yeo, R. (eds) (1996). *Telling Lives in Science: Essays on Scientific Biography*. Cambridge: Cambridge University Press.

Slotten, R. A. (2004). *The Heretic in Darwin's Court: The Life of Alfred Russel Wallace*. New York: Columbia University Press.

Smiles, S. (1859). *Self-help: With Illustrations of Character and Conduct*. London: John Murray.

Smiles, S. (1887). *Life and Labour; or, Characteristics of Men of Industry, Culture and Genius.* London: John Murray.

Smith, C. and Norton Wise, M. (1989). *Energy and Empire: A Biographical Study of Lord Kelvin.* Cambridge: Cambridge University Press.

Smocovitis, V. B. (1999). The 1959 Darwin centennial celebration in America. *Osiris,* **14,** 274–323.

Söderqvist, T. (ed.) (2007). *The History and Poetics of Scientific Biography.* Aldershot, England; Burlington, VT: Ashgate.

Stauffer, R. C. (ed.) (1975). *Charles Darwin's Natural Selection, being the second part of his big species book written from 1856 to 1858.* Cambridge: Cambridge University Press.

Sulloway, F. J. (1982). Darwin and his finches: the evolution of a legend. *Journal of the History of Biology,* **15,** 1–53.

Travers, T. (1987). *Samuel Smiles and the Victorian Work Ethic.* New York: Garland.

White, P. (2003). *Thomas Huxley: Making the 'Man of Science'.* Cambridge: Cambridge University Press.

Winsor, M. P. (2006). The creation of the essentialism story: an exercise in metahistory. *History and Philosophy of the Life Sciences,* **28,** 149–74.

Woodberry, G. E. (1890). *Studies in Letters and Life.* Boston: Houghton Mifflin and Company, p. 253.

Yeo, R. (1985). An idol of the market-place: Baconianism in nineteenth century Britain. *History of Science,* **23,** 251–98.

Yeo, R. (1988). Genius, method and morality: images of Newton in Britain, 1760–1860. *Science in Context,* **2,** 257–84.

Young, R. M. (1988). Darwin and the genre of biography. In *One Culture: Essays in Science and Literature,* ed. George Levine. Madison: University of Wisconsin Press, pp. 203–24.

REFERENCES (JAMES A. SECORD – CHAPTER 2)

Browne, J. (1995). *Charles Darwin: Voyaging.* London: Jonathan Cape, p. 511.

Burkhardt, F., Smith, S. *et al.* (eds) (1985–) *The Correspondence of Charles Darwin.* vols 1–17 and continuing. Cambridge: Cambridge University Press. http://www.darwinproject.ac.uk.

Darnton, R. (1996). *The Forbidden Best-sellers of Pre-revolutionary France.* London: HarperCollins.

Darwin, C. (1871). *The Descent of Man, and Selection in Relation to Sex.* 2 vols. London: John Murray, **2,** p. 404.

Daum, A. W. (2002). *Wissenshafts-popularisierung im 19. Jahrhundert: Bürgerliche Kultur, naturwissenschaftliche Bildung und die deutsche Öffentlichkeit, 1848–1914.* München: R. Oldenbourg, pp. 300–23.

Dewey, J. (1909). Darwin's influence upon philosophy. *Popular Science Monthly,* **75**, 90–98, at p. 98.

Ellegård, A. (1958). *Darwin and the General Reader: The Reception of Darwin's Theory of Evolution in the British Periodical Press, 1859–1872.* Göteborg: Göteborgs Universititets Årsskrift.

Elshakry, M. (2007). The gospel of science and American evangelism in late Ottoman Beirut. *Past and Present,* **196**, 173–214.

Elshakry, M. S. (2008). Knowledge in motion: the cultural politics of modern science translations in Arabic. *Isis,* **99**, 701–30, at p. 714.

Glaß, D. (1994). Popularizing sciences through Arabic journals in the late nineteenth century: how al-Muqtataf transformed western patterns. In *Changing Identities: The Transformation of Asian and African Societies under Colonialism,* ed. J. Heidrich. Berlin: Verlag Das Arabishe Buch, pp. 323–64.

Glick, T. (ed.) (1974). *The Comparative Reception of Darwinism.* Austin: University of Texas Press.

Herschel, J. (1830). *A Preliminary Discourse on the Study of Natural Philosophy.* London: Longman, p. 201.

Huxley, T. H. (1860). Darwin on the origin of species. *Westminster Review,* **n. s. 17**, 541–70, at p. 556.

Mayr, E. (1985). Darwin's five theories of evolution. In *The Darwinian Heritage,* ed. D. Kohn. Princeton: Princeton University Press, pp. 755–72.

Mudie, R. (1838). *Man, in his Physical Structure and Adaptations.* London: W. S. Orr, p. 251.

Nichol, J. P. (1837). Views of the architecture of the heavens. In *A Series of Letters to a Lady.* Edinburgh: William Tait, pp. 206–7.

Secord, J. A. (2000). *Victorian Sensation: The Extraordinary Publication, Reception, and Secret Authorship of Vestiges of the Natural History of Creation.* Chicago: University of Chicago Press.

BIBLIOGRAPHY

Bayly, C. A. (2004). *The Birth of the Modern World, 1780–1914.* Oxford: Blackwell.

Beer, G. (2009). *Darwin's Plots: Evolutionary Narrative in Darwin, George Eliot and Nineteenth-century Fiction,* 3rd edn. Cambridge: Cambridge University Press.

Bowler, P. (2003). *Evolution: The History of an Idea*, 3rd edn. Berkeley and Los Angeles: University of California Press.

Darwin, C. (2008). *Evolutionary Writings*, ed. J. A. Secord. Oxford: Oxford University Press.

Engels, E. M. and Glick, T. F. (eds) (2009). *The Reception of Charles Darwin in Europe*, 2 vols. London: Continuum.

Glick, T., Puig-Samper, M. A., Ruiz, R. (eds) (2001). *The Reception of Darwinism in the Iberian World: Spain, Spanish America and Brazil.* Dordrecht: Kluwer.

Numbers, R. L. and Stenhouse, J. (eds) (1999). *Disseminating Darwinism: The Role of Place, Race, Religion and Gender.* Cambridge: Cambridge University Press.

Secord, J. A. (2004). Knowledge in transit. *Isis*, **95**, 654–72.

Winseck, D. R. and Pike, R. M. (2007). *Communication and Empire: Media, Markets, and Globalization, 1860–1930.* Durham and London: Duke University Press.

REFERENCES (REBECCA STOTT – CHAPTER 3)

Bishop, E. (1964). *Letter to Anne Stevenson. 8–20 Jan.* Washington University, St Louis: Elizabeth Bishop Papers.

de Beer, G. (1974). *Charles Darwin and Thomas Henry Huxley: Autobiographies.* London, New York: OUP.

Beer, G. (1983). *Darwin's Plots: Evolutionary Narrative in Darwin, George Eliot and Nineteenth-Century Fiction.* London: Routledge.

Byatt, A. S. (1990). *Possession.* London: Chatto and Windus.

Carroll, L. (2003). *Alice's Adventures in Wonderland and Through the Looking Glass.* London: Penguin Classic.

Clampitt, A. (1998). *The Collected Poems of Amy Clampitt.* London: Faber.

Darwin, C. (1985). *The Origin of Species,* 1859. Reprinted London: Penguin.

Darwin, C., Burkhardt, F., Smith, S. (1985). *The Correspondence of Charles Darwin.* Cambridge: CUP.

Eliot, G. (1872). *Middlemarch.* London: Penguin.

Kingsley, C. (1864). *The Water-Babies.* Boston: T.O.H.P. Burnham.

McEwan, I. (2005). *Saturday.* London: Jonathan Cape.

Stott, R. (2003). *Darwin and the Barnacle.* London: Faber.

Stott, R. (2007). *Ghostwalk.* London: Weidenfeld and Nicolson.

Stott, R. (2009). *The Coral Thief.* London: Weidenfeld and Nicolson.

Wells, H. G. (1958). The time machine, 1895. In H. G. Wells, *Selected Short Stories.* London: Penguin.

Wells, H. G. (1975). *The War of the Worlds,* 1897. Reprinted London: Pan Books.

BIBLIOGRAPHY

Beer, G. (1986). Darwin's reading and the fictions of development. In *The Darwinian Heritage*, ed. David Kohn. Princeton: Princeton.

Edward William Cooke (1872). *Entwickelungsgeschichte*. London: Longmans, Green.

Foucault, M. (1990). *The History of Sexuality Volume 1: An Introduction*. London: Vintage.

Stott, R. (2000). Through a glass darkly: aquarium colonies and nineteenth-century narratives of marine monstrosity. *Gothic Studies*, **2**(3), 305–27.

REFERENCES (PAUL SEABRIGHT – CHAPTER 4)

Arnqvist, G. and Rowe, L. (2005). *Sexual Conflict*. Princeton: Princeton University Press.

Boehm, C. (1999). *Hierarchy in the Forest: The Evolution of Egalitarian Behavior*. Cambridge, MA: Harvard University Press.

Bowles, S. (2009). Did warfare among ancestral hunter–gatherers affect the evolution of human social behaviors? *Science*, **324**, 1293–8.

Coyne, J. A. (2009). *Why Evolution is True*. New York: Viking Penguin.

Darwin, C. (1859). *On the Origin of Species by Means of Natural Selection*. London: John Murray, facsimile edition published in 1979 by Gramercy Books, a division of Random House, New York.

Darwin, C. (1871). *The Descent of Man, and Selection in Relation to Sex*. London: John Murray, facsimile edition published in 1981 by Princeton University Press.

Desmond, A. and Moore, J. (2009). *Darwin's Sacred Cause: Race, Slavery and the Quest for Human Origins*. London: Allen Lane.

Eisner, M. (2001). Modernization, self-control and lethal violence: the long-term dynamics of European homicide rates in theoretical perspective. *British Journal of Criminology*, **41**, 618–38.

Friebel, G. and Seabright, P. (2009). Evidence from telephone use of gender-dependent preferences for strong versus weak social ties. *Toulouse School of Economics*, mimeo.

Gomes, C. and Boesch, C. (2009). Wild chimpanzees exchange meat for sex on a long-term basis. *PLoS ONE*, **4**(4): e5116. Doi:10.1371/journal.pone.0005116.

Granovetter, M. (1973). The strength of weak ties. *American Journal of Sociology,* **78**, 1360–80.

Kaplan, H., Hill, K., Lancaster, J., Hurtado, A. M. (2000). A theory of human life history evolution: diet, intelligence, and longevity. *Evolutionary Anthropology,* **9**, 156–85.

Kaplan, H., Hooper, P., Gurven, M. (2009). The evolutionary and ecological roots of human social organization. *Philosophical Transactions of the Royal Society B,* **364**, 3289–99.

Keeley, L. (1996). *War Before Civilization: The Myth of the Peaceful Savage.* Oxford: Oxford University Press.

Marmot, M. (2004). *The Status Syndrome: How Social Standing Affects Our Health and Longevity.* New York: Times Books.

Muller, M. and Wrangham, R. (2004). Dominance, cortisol and stress in wild chimpanzees (*Pan troglodytes schweinfurthii*). *Behavioral Ecology and Sociobiology,* **55**, 332–40.

Pinker, S. (1994). *The Language Instinct.* New York: William Morrow & Co.

Ridley, M. (2004). *Evolution,* 3rd edn. Oxford: Blackwell.

Roughgarden, J. (2009). *The Genial Gene: Deconstructing Darwinian Selfishness.* Berkeley: University of California Press.

Stanford, C. B. (1999). *The Hunting Apes: Meat-Eating and the Origins of Human Behavior.* Princeton: Princeton University Press.

de Waal, F. (1982). *Chimpanzee Politics: Power and Sex among Apes.* New York: Harper & Row.

de Waal, F. (1989). *Peacemaking Among Primates.* Cambridge, MA: Harvard University Press.

de Waal, F. (ed.) (2001). *Tree of Origin: What Primate Behavior Can Tell Us About Human Social Evolution.* Cambridge, MA: Harvard University Press.

Wrangham, R. and Peterson, D. (1996). *Demonic Males: Apes and the Origins of Human Violence.* New York: Houghton Mifflin.

BIBLIOGRAPHY (STEVE JONES – CHAPTER 5)

Crow, J. F. (2006). Age and sex effects on human mutation: an old problem with new complexities. *Journal of Radiation Research,* **47** (Suppl.), B75–B82.

Keinan, A., Mullikin, J. C., Patterson, N., Reich, D. (2008). Accelerated genetic drift on chromosome X during the human dispersal out of Africa. *Nature Genetics,* **41**, 66–70.

Levy, S., Sutton, G., Ng, P. C. *et al.* (2007). The diploid genome sequence of an individual human. *PLoS Biology,* **5**(10), e 254. doi:10.1371/journal.pbio.0050254.

Li, W. H., Yi, S. J., Makova, K. (2002). Male-driven evolution. *Current Opinion in Genetics & Development*, **12**, 650–6.

McQuillan, R., Leutenegger, A., Abdel-Rahman, R. *et al.* (2008). Runs of homozygosity in European populations. *American Journal of Human Genetics*, **83**, 359–72.

Moore, L. T., McEvoy, B., Cape, E., Simms, K., Bradley, D. G. (2006). A Y-chromosome signature of hegemony in Gaelic Ireland. *American Journal of Human Genetics*, **78**, 334–8.

Nalls, M. A., Simon-Sanchez, J., Gibbs, J. R. *et al.* (2009). Measures of autozygosity in decline: globalization, urbanization, and its implications for medical genetics. *PLOS Genetics*, **5**, e1000415.

Parra, E. J. (2007). Human pigment variation: evolution, genetic basis, and implications for public health. *American Journal of Physical Anthropology*, S45, 85–105.

Pinhasi, R., Fort, J., Ammerman, A. J. (2005). Tracing the origin and spread of agriculture. *PLoS Biol*, **3**(12), e410.

Richardson, D. B. (2009). Exposure to ionizing radiation and thyroid cancer incidence. *Epidemiology*, **20**, 181–7.

Tishkoff, S. A., Reed, F. A., Friedlaender, F. R. *et al.* (2009). The genetic structure and history of Africans and African Americans. *Science*, **324**, 1035–44.

REFERENCES (SEAN B. CARROLL – CHAPTER 6)

Carroll, S. B. (2006). *The Making of the Fittest: DNA and the Ultimate Forensic Record of Evolution*. New York: W. W. Norton.

Chen, L., DeVries, A. L., Cheng, C -H. C. (1997). Evolution of antifreeze glycoprotein gene from a trypsinogen gene in Antarctic notothenioid fish. *Proceedings of the National Academy of Sciences, USA*, **94**, 3811–16.

Cheng, C -H. C. and Chen, L. (1999). Evolution of an antifreeze glycoprotein. *Nature*, **401**, 443–4.

Cocca, E., Ratnayake-Lecamwasam, M., Parker, S. K. *et al.* (1995). Genomic remnants of α-globin in the hemoglobinless Antarctic icefishes. *Proceedings of the National Academy of Sciences, USA*, **92**, 1817–21.

Darwin, C. (1859). *On the Origin of Species by Means of Natural Selection*. London: John Murray, facsimile edition published in 1979 by Gramercy Books, a division of Random House, New York.

David-Gray, Z. K., Bellingham, J., Munoz, M. *et al.* (2002). Adaptive loss of ultraviolet-sensitive/violet-sensitive (UVS/VS) cone opsin in the blind mole rat (*Spalax ehrenbergi*). *European Journal of Neuroscience*, **16**, 1186–94.

DeVries, A. L. (1971). Glycoproteins as biological antifreeze agents in Antarctic fishes. *Science,* **172**, 1152–5.

DeVries, A. L. and Wohlschlag, D. E. (1969). Freezing resistance in some Antarctic fishes. *Science,* **163**, 1073–5.

Jacobs, G. H., Neitz, M., Neitz, J. (1996). Mutations in S-cone pigment genes and the absence of colour vision in two species of nocturnal primate. *Proceedings of the Royal Society of London B,* **263**, 705–10.

Nachman, M. W., Hoekstra, H. E., D'Agostino, S. L. (2003). The genetic basis of adaptive melanism in pocket mice. *Proceedings of the National Academy of Sciences, USA,* **100**, 5268–73.

Norske Videnskaps-Akademi i Oslo (1935). In *Scientific Results of the Norwegian Antarctic Expeditions, 1927–1928,* ed. Olaf Holtedahl. Oslo: I kommisjon hos J. Dybwad.

Ödeen, A. and Håstad, O. (2003). Complex distribution of avian color vision systems revealed by sequencing the SWS1 opsin from total DNA. *Molecular Biology and Evolution,* **20**, 855–61.

Ruud, J. T. (1954). Vertebrates without erythrocytes and blood pigment. *Nature,* **173**, 848–50.

Ruud, J. T. (1965). The ice fish. *Scientific American,* **213**, 108–15.

Shi, Y. and Yokoyama, S. (2003). Molecular analysis of the evolutionary significance of ultraviolet vision in vertebrates. *Proceedings of the National Academy of Sciences, USA,* **100**, 8308–13.

Somero, G. N. and DeVries, A. L. (1967). Temperature tolerance of some Antarctic fishes. *Science,* **156**, 257–8.

Viitala, J., Korplmaki, E., Palokangas, P., Koivula, M. (1995). Attraction of kestrels to vole scent marks visible in ultraviolet light. *Nature,* **373**, 425–7.

Yokoyama, S., Huan Zhang, F., Radlwimmer, B., Blow, N. S. (1999). Adaptive evolution of color vision of the Comoran coelacanth (*Latimeria chalumnae*). *Proceedings of the National Academy of Sciences, USA,* **96**, 6279–84.

REFERENCES (CRAIG MORITZ & ANA CAROLINA CARNAVAL – CHAPTER 7)

Arnold, M. L. (1997). *Natural Hybridization and Evolution.* New York: Oxford University Press.

Avise, J. C. (2004). *Molecular Markers, Natural History and Evolution.* Massachusetts: Sinauer Associates.

Bell, K., Moritz, C., Moussalli, A., Yeates, D. (2007). Comparative phylogeography and speciation of dung beetles from the Australian wet tropics rainforest. *Molecular Ecology,* **16**, 4984–98.

Bossuyt, F., Meegaskumbura, M., Beenaerts, N. *et al.* (2004). Local endemism within the western ghats-Sri Lanka biodiversity hotspot. *Science*, **306**, 479–81.

Bowman, D. M. J. S., Balch, J. K., Artaxo, P. *et al.* (2009). Fire in the Earth system. *Science*, **324**, 481–4.

Brooks, T. M., Mittermeier, R. A., da Fonseca, G. A. B. *et al.* (2006). Global biodiversity conservation priorities. *Science*, **313**, 58–61.

Carnaval, A. C. and Moritz, C. (2008). Historical climate modelling predicts patterns of current biodiversity in the Brazilian Atlantic forest. *Journal of Biogeography*, **35**, 1187–1201.

Carnaval, A. C., Hickerson, M. J., Haddad, C. F. B., Rodrigues, M. T., Moritz, C. (2009). Stability predicts genetic diversity in the Brazilian Atlantic forest hotspot. *Science*, **323**, 785–9.

Carroll, S. P., Hendry, A. P., Reznick, D. N., Fox, C. W. (2007). Evolution on ecological time-scales. *Functional Ecology*, **21**, 387–93.

Clegg, S. M., Degnan, S. M., Kikkawa, J. *et al.* (2002). Genetic consequences of sequential founder events by an island-colonizing bird. *Proceedings of the National Academy of Sciences, USA*, **99**, 8127–32.

Colinvaux, P. A., De Oliveira, P. E., Bush, M. (2000). Amazonian and neotropical plant communities on glacial time-scales: the failure of the aridity and refuge hypothesis. *Quaternary Science Reviews*, **19**, 141–69.

Colwell, R. K., Brehm, G., Cardelus, C. L., Gilman, A. C., Longino, J. T. (2008). Global warming, elevational range shifts, and lowland biotic attrition in the wet tropics. *Science*, **322**, 258–261. DOI: 10.1126/science.1162547.

Coyne, J. A., Barton, N. H., Turelli, M. (1997). Perspective: a critique of Sewall Wright's shifting balance theory of evolution. *Evolution*, **51**, 643–71.

Crow, J. F. (2008). Mid-century controversies in population genetics. *Annual Review of Genetics*, **42**, 1–16.

Darwin, C. R. (1842). *The Structure and Distribution of Coral Reefs*. London: Smith Elder and Co.

Darwin, C. R. (1859). *On the Origin of Species by Means of Natural Selection, or the Preservation of Favoured Races in the Struggle for Life*. London: John Murray.

Davis, E. B., Koo, M. S., Conroy, C., Patton, J. L., Moritz, C. (2008). The California Hotspots Project: identifying regions of rapid diversification of mammals. *Molecular Ecology*, **17**, 120–38.

Deutsch, C. A., Tewksbury, J. J., Huey, R. B. *et al.* (2008). Impacts of climate warming on terrestrial ectotherms across latitude. *Proceedings of the National Academy of Sciences, USA,* **105**, 6668–72.

Donoghue, M. J. (2008). A phylogenetic perspective on the distribution of plant diversity. In *In the Light of Evolution.* Vol. II: *Biodiversity and Extinction,* eds J. C. Avise *et al.* Washington, DC: Proceedings of the National Academy of Sciences of the United States of America.

Fjeldså, J. and Lovett, J. C. (1997). Geographical patterns of old and young species in African forest biota: the significance of specific montane areas as evolutionary centres. *Biodiversity and Conservation,* **6**, 325–46.

Frankel, O. (1974). Genetic conservation: our evolutionary responsibility. *Genetics,* **78**, 53–65.

Funk, V. A. and Wagner, W. L. (1995). Biogeographic patterns in the Hawaiian islands. In *Hawaiian Biogeography: Evolution on a Hot Spot Archipelago,* eds W. L. Wagner and V. A. Funk. Washington DC: Smithsonian Institution Press, pp. 379–419.

Gascon, C., Malcolm, J. R., Patton, J. L. *et al.* (2000). Riverine barriers and the geographic distribution of Amazonian species. *Proceedings of the National Academy of Sciences, USA,* **97**, 13672–7.

Gavrilets, S. (2003). Models of speciation: what have we learned in 40 years? *Evolution,* **57**, 2197–215.

Ghalambor, C. K., Huey, R. B., Martin, P. R., Tewksbury, J. J., Wang, G. (2006). Are mountain passes higher in the tropics? Janzen's hypothesis revisited. *Integrative and Comparative Biology,* **46**, 5–17.

Gillespie, R. G. (2004). Community assembly through adaptive radiation in Hawaiian spiders. *Science,* **303**, 356–9.

Gillespie, R. G. and Baldwin, B. G. (2009). Island biogeography of remote archipelagos: interplay between ecological and evolutionary processes. In *In The Theory of Island Biogeography at 40: Impacts and Prospects,* eds J. B. Losos and R. E. Ricklefs. Princeton: Princeton University Press.

Gillespie, R. G. and Roderick, G. K. (2002). Arthropods on islands: evolution and conservation. *Annual Review of Entomology,* **47**, 595–632.

Gillespie, R. G., Claridge, E. M., Goodacre, S. (2008a). Biogeography of French Polynesia: diversification within and between a series of hotspot archipelagoes. *Philosophical Transactions Royal Society London,* **363**, 3335–46.

Gillespie, R. G., Claridge, E. M., Roderick, G. K. (2008b). Biodiversity dynamics in isolated island communities: interaction between natural and human-mediated processes. *Molecular Ecology*, 17, 45–57.

Givnish, T. J., Millam, K. C., Mast, A. R. *et al.* (2009). Origin, adaptive radiation and diversification of the Hawaiian lobeliads (Asterales: Campanulaceae). *Proceedings of the Royal Society B: Biological Sciences*, **276**, 407–16.

Goldberg, J., Trewick, S. A., Paterson, A. M. (2008). Evolution of New Zealand's terrestrial fauna: a review of molecular evidence. *Philosophical Transactions of the Royal Society B: Biological Sciences*, **363**, 3319–34.

Graham, C. H., Ferrier, S., Huettman, F., Moritz, C., Peterson, A. T. (2004). New developments in museum-based informatics and applications in biodiversity analysis. *Trends in Ecology & Evolution*, **19**, 497–503.

Graham, C. H., Moritz, C., Williams, S. E. (2006). Habitat history improves prediction of biodiversity in rainforest fauna. *Proceedings of the National Academy of Sciences, USA*, **103**, 632–6.

Grant, B. R. and Grant, P. R. (1996). High survival of Darwin's finch hybrids: effects of beak morphology and diets. *Ecology*, **77**, 500–9.

Heaney, L. R. (2007). Is a new paradigm emerging for oceanic island biogeography? *Journal of Biogeography*, **34**, 753–7.

Hewitt, G. M. (2004). Genetic consequences of climatic oscillations in the Quaternary. *Philosophical Transactions of the Royal Society of London. Series B: Biological Sciences*, **359**, 183–95.

Hey, J. and Machado, C. A. (2003). The study of structured populations – new hope for a difficult and divided science. *Nature Reviews Genetics*, **4**, 535–43.

Hickerson, M. J., Stahl, E. A., Lessios, H. A., Crandall, K. (2006). Test for simultaneous divergence using approximate Bayesian computation. *Evolution*, **60**, 2435–53.

Hooker, J. D. (1844). *Flora Antartica: The Botany of the Antarctic Voyage 1839–1843*. London: Reeve.

Hugall, A., Moritz, C., Moussalli, A., Stanisic, J. (2002). Reconciling paleodistribution models and comparative phylogeography in the wet tropics rainforest land snail *Gnarosophia bellendenkerensis* (Brazier 1875). *Proceedings of the National Academy of Sciences, USA*, **99**, 6112–17.

Hughes, C. and Eastwood, R. (2006). Island radiation on a continental scale: exceptional rates of plant diversification after uplift of the

Andes. *Proceedings of the National Academy of Sciences,* **103**, 10334–9. DOI: 10.1073/pnas.0601928103.

Jansson, R. and Dynesius, M. (2002). The fate of clades in a world of recurrent climatic change: Milankovitch oscillations and evolution. *Annual Review of Ecology and Systematics,* **33**, 741–77.

Janzen, D. H. (1967). Why mountain passes are higher in the tropics. *American Naturalist,* **101**, 233–49.

Jetz, W. and Rahbek, C. (2002). Geographic range size and determinants of avian species richness. *Science,* **297**, 1548–51.

Kahindo, C., Bowie, R. C. K., Bates, J. M. (2007). The relevance of data on genetic diversity for the conservation of Afro-montane regions. *Biological Conservation,* **134**, 262–70.

Kirkpatrick, M. and Ravigné, V. (2002). Speciation by natural and sexual selection: models and experiments. *The American Naturalist,* **159**, S22–S35.

Koch, P. L. and Barnosky, A. D. (2006). Late Quaternary extinctions: state of the debate. *Annual Review of Ecology, Evolution, and Systematics,* **37**, 215–50.

Kozak, K. H. and Wiens, J. J. (2007). Climatic zonation drives latitudinal variation in speciation mechanisms. *Proceedings of the Royal Society B: Biological Sciences,* **274**, 2995–3003.

Kremen, C., Cameron, A., Moilanen A. *et al.* (2008). Aligning conservation priorities across taxa in Madagascar with high-resolution planning tools. *Science,* **320**, 222–6.

Lemmon, A. R. and Lemmon, E. M. (2008). A likelihood framework for estimating phylogeographic history on a continuous landscape. *Systematic Biology,* **57**, 544–61.

Lessa, E. P., Cook, J. A., Patton, J. L. (2003). Genetic footprints of demographic expansion in North America, but not Amazonia, during the late Quaternary. *Proceedings of The National Academy of Sciences, USA,* **100**, 10331–4.

Lomolino, M. V., Sax, D. F., Brown, J. H. (2004). *Foundations of Biogeography: Classic Papers with Commentaries.* Chicago: University of Chicago Press.

Losos, J. B. (2007). Detective work in the West Indies: integrating historical and experimental approaches to study island lizard evolution. *Bioscience,* **57**, 585–97.

Lovette, I. J., Bermingham, E., Ricklefs, R. E. (2002). Clade-specific morphological diversification and adaptive radiation in Hawaiian songbirds. *Proceedings of the Royal Society of London, Series B,* **269**, 37–42.

Lyell, C. (1830). *Principles of Geology*. London: John Murray.

Martínez-Meyer, E., Peterson, A. T., Hargrove, W. W. (2004). Ecological niches as stable distributional constraints on mammal species, with implications for Pleistocene extinctions and climate change projections for biodiversity. *Global Ecology & Biogeography*, **13**, 305–14.

Moritz, C. (2002). Strategies to protect biological diversity and the processes that sustain it. *Systematic Biology*, **51**, 238–54.

Moritz, C., Hoskin, C. J., MacKenzie, J. B. *et al.* (2009). Identification and dynamics of a cryptic suture zone in tropical rainforest. *Proceedings of the Royal Society B: Biological Sciences*, **276**, 1235–44.

Moritz, C., Patton, J. L., Schneider, C. J., Smith, T. B. (2000). Diversification of rainforest faunas: an integrated molecular approach. *Annual Review of Ecology and Systematics*, **31**, 533–63.

Moritz, C., Richardson, K. S., Ferrier, S. *et al.* (2001). Biogeographical concordance and efficiency of taxon indicators for establishing conservation priority in a tropical rainforest biota. *Proceedings of the Royal Society London, Biological Sciences*, **268**, 1–7.

Moussalli, A., Moritz, C., Williams, S. E., Carnaval, A. C. (2009). Variable responses of skinks to a common history of rainforest fluctuation: concordance between phylogeography and palaeo-distribution models. *Molecular Ecology*, **18**, 483–99.

Nielsen, R. and Beaumont, M. A. (2009). Statistical inferences in phylogeography. *Molecular Ecology*, **18**, 1034–47.

Nix, H. A. (1991). Biogeography: patterns and process. In *Rainforest Animals. Atlas of Vertebrates Endemic to Australia's Wet Tropics*, eds H. A. Nix and M. Switzer. Canberra: Australian Nature Conservation Agency, pp. 11–40.

Orr, H. A. and Orr, L. H. (1996). Waiting for speciation: the effect of population subdivision on the waiting time to speciation. *Evolution*, **50**, 1742–9.

Parent, C. E., Caccone, A., Petren, K. (2008). Colonization and diversification of Galápagos terrestrial fauna: a phylogenetic and biogeographical synthesis. *Philosophical Transactions of the Royal Society B: Biological Sciences*, **363**, 3347–61.

Patton, J. L., Da Silva, M. N., Malcolm, J. R. (2000). Mammals of the Rio Juruá and the evolutionary and ecological diversification of Amazonia. *Bulletin of the American Museum of Natural History*, **244**, 1–306.

Petit, R. J., Brewer, S., Bordács, S. *et al.* (2002). Identification of refugia and post-glacial colonisation routes of European white oaks based on chloroplast DNA and fossil pollen evidence. *Forest Ecology and Management,* **156,** 49–74.

Price, J. P. and Clague, D. A. (2002). How old is the Hawaiian biota? Geology and phylogeny suggest recent divergence. *Proceedings of the Royal Society of London,* **269,** 2429–35.

Provine, W. (2004). Speciation in historical perspective. In *Adaptive Speciation,* eds U. Dieckmann *et al.* Cambridge: Cambridge University Press, pp. 17–29.

Rahbek, C., Gotelli, N. J., Colwell, R. K. *et al.* (2007). Predicting continental-scale patterns of bird species richness with spatially explicit models. *Proceedings of the Royal Society B: Biological Sciences,* **274,** 165–74.

Reding, D. M., Foster, J. T., James, H. F., Pratt, H. D., Fleischer, R. C. (2009). Convergent evolution of 'creepers' in the Hawaiian honeycreeper radiation. *Biology Letters,* **5,** 221–4.

Richards, C. L., Carstens, B. C., Knowles, L. L. (2007). Distribution modelling and statistical phylogeography: an integrative framework for generating and testing alternative biogeographical hypotheses. *Journal of Biogeography,* **34,** 1833–45.

Roderick, G. K. and Percy, D. M. (in press). Host plant use, diversification, and coevolution: insights from remote oceanic islands. In *Evolutionary Biology of Plant and Insect Relationships,* ed. K. J. Tilmon. Berkeley: University of California Press.

Rull, V. (2005). Biotic diversification in the Guayana Highlands: a proposal. *Journal of Biogeography,* **32,** 921–7.

Sanmartín, I. and Ronquist, F. (2004). Southern hemisphere biogeography inferred by event-based models: plant vs. animal patterns. *Systematic Biology,* **53,** 216–43.

Schluter, D. (2001). Ecology and the origin of species. *Trends in Ecology and Evolution,* **16,** 372–80.

Schneider, C. J., Cunningham, C., Moritz, C. (1998). Comparative phylogeography and the history of endemic vertebrates in the wet tropics rainforests of Australia. *Molecular Ecology,* **7,** 487–98.

Simon, F. (2002). Mapping spatial pattern in biodiversity for regional conservation planning: where to from here? *Systematic Biology,* **51,** 331–63.

Steadman, D. W. and Martin, P. S. (2003). The late Quaternary extinction and future resurrection of birds on Pacific islands. *Earth Science Reviews,* **61,** 133–47.

Svenning, J. -C., Normand, S., Skov, F. (2008). Postglacial dispersal limitation of widespread forest plant species in nemoral Europe. *Ecography,* **31,** 316–26.

Thompson, J. N. (2009). The coevolving web of life. *American Naturalist*, **173**, 125–40.

Uyeda, J. C., Arnold, S. J., Hohenlohe, P. A., Mead, L. S., Kokko, H. (2009). Drift promotes speciation by sexual selection. *Evolution*, **63**, 583–94.

VanDerWal, J., Shoo, L. P., Williams, S. E. (2009). New approaches to understanding late Quaternary climate fluctuations and refugial dynamics in Australian wet tropical rain forests. *Journal of Biogeography*, **36**, 291–301.

Vieites, D. R., Wollenberg, K. C., Andreone, F. *et al.* (2009). Vast underestimation of Madagascar's biodiversity evidenced by an integrative amphibian inventory. *Proceedings of the National Academy of Sciences*, **106**, 8267–72.

Wallace, A. R. (1852). On the monkeys of the Amazon. *Proceedings of the Zoological Society of London*, **20**, 107–10.

Wallace, A. R. (1855). On the law which has regulated the introduction of new species. *Annals and Magazine of Natural History*, **16**, (2nd series) 184–96.

Wallace, A. R. (1860). On the zoological geography of the Malay archipelago. *Journal of the Proceedings of the Linnean Society*, **4**, 172–84.

Wallace, A. R. (1865). On the phenomena of variation and geographical distribution as illustrated by the Papilionidae of the Malayan region. *Transactions of the Linnean Society of London*, **25**, 1–71.

Waltari, E., Hoberg, E. P., Lessa, E. P., Cook, J. A. (2007). Eastward Ho: phylogeographical perspectives on colonization of hosts and parasites across the Beringian nexus. *Journal of Biogeography*, **34**, 561–74.

Wegener, A. (1915). *Die Entstehung der Kontinente und Ozeane Friedrich Vieweg & Sons.* Braunschweig: Friedr. Vieweg & Sohn Akt-Ges.

Whittaker, R. J., Araújo, M. B., Jepson, P. *et al.* (2005). Conservation biogeography: assessment and prospect. *Diversity and Distributions*, **11**, 3–23.

Whittaker, R. J. and Fernandez-Palacios, J. M. (2007). *Island Biogeography – Ecology, Evolution, and Conservation.* Oxford: Oxford University Press.

Whittaker, R. J., Triantis, K. A., Ladle, R. J. (2008). A general dynamic theory of oceanic island biogeography. *Journal of Biogeography*, **35**, 977–94.

Wiens, J. J., Parra-Olea, G., García-París, M., Wake, D. B. (2007). Phylogenetic history underlies elevational biodiversity patterns in tropical salamanders. *Proceedings of the Royal Society B: Biological Sciences,* **274**, 919–28.

Williams, J. W., Jackson, S. T., Kutzbach, J. E. (2007). Projected distributions of novel and disappearing climates by 2100 AD. *Proceedings of the National Academy of Sciences,* **104**, 5738–42.

Williams, S. E., Bolitho, E. E., Fox, S. (2003). Climate change in Australian tropical rainforests: an impending environmental catastrophe. *Proceedings of the Royal Society of London. Series B: Biological Sciences,* **270**, 1887–92.

REFERENCES (JOHN DUPRÉ – CHAPTER 8)

Barnes, B. and Dupré, J. (2008). *Genomes and What to Make of Them.* Chicago: University of Chicago Press.

Bates, J. M., Mittgeb, E., Kuhlman, J. *et al.* (2006). Distinct signals from the microbiota promote different aspects of zebrafish gut differentiation. *Developmental Biology,* **297**, 374–86.

Boucher, Y., Douady, C. J., Papke, R. T. *et al.* (2003). Lateral gene transfer and the origins of prokaryotic groups. *Annual Review of Genetics,* **37**, 283–328.

Buss, L. (1987). *The Evolution of Individuality.* Princeton, NJ: Princeton University Press.

Charlebois, R. L. and Doolittle, W. F. (2004). Computing prokaryotic gene ubiquity: rescuing the core from extinction. *Genome Research,* **14**, 2469–77.

Chong, S. and Whitelaw, E. (2004). Epigenetic germline inheritance. *Current Opinion in Genetics & Development,* **14**, 692–6.

Dawkins, R. (1976). *The Selfish Gene.* Oxford: Oxford University Press.

Dennett, D. (1995). *Darwin's Dangerous Idea: Evolution and the Meanings of Life.* New York: Simon & Schuster.

Doolittle, F. (1999). Phylogenetic classification and the universal tree. *Science,* **284**, 2124–9.

Dunning Hotopp, J. C., Clark, M. E., Oliveira, D. C. *et al.* (2007). Widespread lateral gene transfer from intracellular bacteria to multicellular eukaryotes. *Science,* **317**, 1753.

Dupré, J. (2001). *Human Nature and the Limits of Science.* Oxford: Oxford University Press.

Dupré, J. (2008). *The Constituents of Life.* Amsterdam: Van Gorcum.

Dupré, J. and O'Malley, M. (2007). Metagenomics and biological ontology. *Studies in the History and Philosophy of the Biological and Biomedical Sciences,* **38**, 834–46.

Dupré, J. and O'Malley, M. (2009). Varieties of living things: life at the intersection of lineage and metabolism. *Philosophy and theory in biology*, 2009 *http://hdl.handle.net/2027/spo.6959004.0001.003.*

Hooper, L. V., Wong, M. H., Thelin, A. *et al.* (2001). Molecular analysis of commensal host–microbial relationships in the intestine. *Science*, **291**, 881–4.

Hughes, W. O. H., Oldroyd, B. P., Beekman, M., Ratnieks, F. L. W. (2008). Ancestral monogamy shows kin selection is key to the evolution of eusociality. *Science*, **320**, 1213–16.

Jablonka, E. and Lamb, M. (1995). *Epigenetic Inheritance and Evolution: The Lamarckian Dimension.* Oxford: Oxford University Press.

Kolenbrander, P. E. (2000). Oral microbial communities: biofilms, interactions, and genetic systems. *Annual Review of Microbiology*, **54**, 413–37.

Krutzen M., Mann, J., Heithaus, M. R. *et al.* (2005). Cultural transmission of tool use in bottlenose dolphins. *Proceedings of the National Academy of Sciences, USA*, **25**, 8939–43.

Lawrence, J. G. and Hendrickson, H. (2005). Genome evolution in bacteria: order beneath chaos. *Current Opinion in Microbiology*, **8**, 572–8.

McFall-Ngai, M. J. (2002). Unseen forces: the influence of bacteria on animal development. *Developmental Biology*, **242**, 1–14.

Mallet, F., Bouton, O., Prudhomme, S. *et al.* (2004). The endogenous retroviral locus ERVWE1 is a bona fide gene involved in hominoid placental physiology. *Proceedings of the National Academy of Sciences, USA*, **101**, 1731–6.

Mallet, J. (2008). Hybridization, ecological races, and the nature of species: empirical evidence for the ease of speciation. *Philosophical Transactions of the Royal Society B-Biological Sciences*, **363**, 2971–86.

Margulis, L. (1970). *Origin of Eukaryotic Cells.* New Haven, CT: Yale University Press.

Margulis, L. and Sagan, D. (2002). *Acquiring Genomes: A Theory of the Origins of Species.* New York: Basic Books.

Marri, P. R., Hao, W., Golding, G. B. (2007). The role of laterally transferred genes in adaptive evolution. *BMC Evolutionary Biology*, **7** (Suppl. 1), S8.

Marsh, P. D. (2004). Dental plaque as a microbial biofilm. *Caries Research*, **38**, 204–11.

Meaney, M. J., Moshe S., Seckl, J. R. (2007). Epigenetic mechanisms of perinatal programming of hypothalamic-pituitary-adrenal function and health. *Trends in Molecular Medicine*, **13**, 269–77.

O'Malley, M. and Dupré, J. (2007). Size doesn't matter: towards a more inclusive philosophy of biology. *Biology and Philosophy,* **22**, 155–91.

Oyama, S. (2000). *The Ontogeny of Information. Developmental Systems and Evolution,* 2nd edn. Durham, NC: Duke University Press.

Oyama, S., Griffiths, P. E., Gray, R. D. (2001). *Cycles of Contingency. Developmental Systems and Evolution.* Cambridge, MA: MIT Press.

Page, R. A. and Ryan, M. J. (2006). Social transmission of novel foraging behavior in bats: frog calls and their referents. *Current Biology,* **16**, 1201–5.

Pal, C., Papp, B., Lercher, M. J. (2005). Adaptive evolution of bacterial metabolic networks by horizontal gene transfer. *Nature Genetics,* **37**, 1372–5.

Powell, A. and Dupré, J. (2009). From molecules to systems: the importance of looking both ways. *Studies in the History and Philosophy of the Biological and Biomedical Sciences,* **40**, 54–64.

Richerson, P. J. and Boyd, R. (2005). *Not By Genes Alone: How Culture Transformed Human Evolution.* Chicago: Chicago University Press.

Savage, D. C. (1977). Microbial ecology of the gastrointestinal tract. *Annual Review of Microbiology,* **31**, 107–33.

Slater, P. J. B. (1986). The cultural transmission of bird song. *Trends in Ecology & Evolution,* **1**, 94–7.

Sober, E. and Sloan Wilson, D. (1998). *Unto Others: The Evolution and Psychology of Unselfish Behavior.* Cambridge, MA: Harvard University Press.

Teixeira L., Ferreira, A., Ashburner, M. (2008). The bacterial symbiont *Wolbachia* induces resistance to RNA viral infections in *Drosophila melanogaster. PLoS Biology,* **6**(12), e2.

Umesaki, Y. and Setoyama, H. (2000). Structure of the intestinal flora responsible for development of the gut immune system in a rodent model. *Microbes and Infection,* **2**, 1343–51.

Weaver, I., Cervoni, N., Champagne, F. A. *et al.* (2004). Epigenetic programming by maternal behaviour. *Nature Neuroscience,* **7**, 847–54.

West Eberhard, M. J. (2003). *Developmental Plasticity and Evolution.* New York: Oxford University Press.

Whitworth, T. L., Dawson, R. D., Magalon, H., Baudry, E. (2007). DNA barcoding cannot reliably identify species of the blowfly genus *Protocalliphora* (Diptera: Calliphoridae). *Proceedings of the Royal Society B,* **274**, 1731–9.

Xu, J. and Gordon, J. I. (2003). Honor thy symbionts. *Proceedings of the National Academy of Sciences, USA,* **100**, 10452–9.

Notes on contributors

Janet Browne is the Aramont Professor of the History of Science at Harvard University. After a first degree in zoology she studied for a PhD in the history of science at Imperial College London, published as *The Secular Ark: Studies in the History of Biogeography* (1983). She previously worked at the Wellcome Trust Centre for the History of Medicine at University College London. She was Associate Editor of the early volumes of *The Correspondence of Charles Darwin*, and more recently as author of a prize-winning biographical study that integrated Darwin's science with his life and times: *Charles Darwin: Voyaging* and *Charles Darwin: The Power of Place*.

Jim A. Secord is Professor of History and Philosophy of Science at the University of Cambridge, and Director of the Darwin Correspondence Project. His research is on the history of science from the late eighteenth to the early twentieth centuries. His publications include *Controversy in Victorian Geology*, editions of the works of Mary Somerville, Charles Lyell and Robert Chambers, and *Victorian Sensation: The Extraordinary Publication, Reception, and Secret Authorship of Vestiges of the Natural History of Creation*.

Rebecca Stott is Professor of Literature and Creative Writing at the University of East Anglia in Norwich where she teaches both creative writing and nineteenth-century literature. She is the author of a number of books and articles about nineteenth-century poets as well as on the cross-fertilisations of literature and science. These include *The Fabrication of the Late Nineteenth-Century Femme Fatale*, *Oyster* and *Darwin and the Barnacle*. As a novelist she is the author of *Ghostwalk* and *The Coral Thief*.

Paul Seabright is Professor of Economics at the University of Toulouse. He was previously a Fellow of All Souls College, Oxford and of Churchill College, Cambridge where he taught in the Economics Faculty. His economic research has been on imperfect competition and the issues it raises for public policy. He is the author of *The Company of Strangers: A Natural History of Economic Life*. He writes regularly for the *London Review of Books*, the *Financial Times* and *Le Monde*.

Steve Jones is Professor of Genetics at University College London, having received his undergraduate and research training from the University of Edinburgh. His work as a molecular biologist has been on, among other topics, genetic variation in snails and fruit flies. He has made a major contribution to the public understanding of science, with books including *Almost Like a Whale – the Origin of Species Updated* and *Coral*. He has been a Reith Lecturer and has been awarded the Royal Society Faraday Medal for public understanding of science.

Sean B. Carroll is Professor of Molecular Biology and Genetics in the Howard Hughes Medical Institute at the University of Wisconsin. His first degree was in biology at Washington University in St Louis; his research degree was in immunology at Tufts Medical School. His work has centred on those genes that control body patterns and play major roles in the evolution of animal diversity. He is the author of *Remarkable Creatures: Epic Adventures in the Search for the Origins of Species*, of *The Making of the Fittest*, and of *Endless Forms Most Beautiful: The New Science of Evo Devo*.

Craig Moritz is Director of the Museum of Vertebrate Zoology and Professor of Integrative Biology at the University of California, Berkeley. He was previously at the University of Queensland. His research applies molecular analysis to the study of ecology and evolution. It explores how species develop and how their populations change in response to environmental change; molecular population genetics of endangered species; the evolution of rainforest fauna; and the mapping of evolutionary hotspots. His books include *Genetics and the Conservation of Wild Populations*.

Ana Carolina Carnaval is a postdoctoral Fellow of the University of California, Berkeley. Her current research projects seek to investigate the effects of tropical rainforest fragmentation on amphibian assemblages in Brazil via a combination of ecological and molecular approaches. She is also pursuing studies about the evolutionary relationships of frog species distributed throughout the Brazilian Atlantic rainforest, hopefully addressing how those relate to Amazonian forms.

John Dupré is currently Professor of Philosophy of Science at the University of Exeter and Director of the ESRC Centre for Genomics in Society (Egenis). He has formerly held posts at Oxford, Stanford and Birkbeck College, London. He is a philosopher of science whose work has focused on issues in biology. His publications include *The Disorder of Things: Metaphysical Foundations of the Disunity of Science*; *Human Nature and the Limits of Science*; *Humans and Other Animals*; *Darwin's Legacy: What Evolution Means Today*; and he has co-authored *Genomes and What to Make of Them* (Chicago, 2008) with the sociologist of science, Barry Barnes.

Index